MOUSETRAPS AND MUFFLING CUPS

ALSO BY KENNETH LASSON

The Workers
Proudly We Hail
Private Lives of Public Servants
Getting the Most Out of Washington
Representing Yourself

MOUSETRAPS

One Hundred Brilliant and Bizarre United States Patents

and MUFFLING CUPS

KENNETH LASSON

Arbor House
New York

Copyright © 1986 by Kenneth Lasson
All rights reserved, including the right of
reproduction in whole or in part in any form.
Published in the United States of America by
Arbor House Publishing Company and in
Canada by Fitzhenry & Whiteside Ltd.

Manufactured in the United States of America

10 9 8 7 6 5 4 3 2 1

Library of Congress Cataloging in Publication Data

Lasson, Kenneth.
   Mousetraps and muffling cups.

  1. Patents—United States.
2. Inventions—United States. I. Title. II. Title:
Mouse traps and muffling cups.
T223.Z1L37   1986   608.773   85-28741
ISBN: 0-87795-786-X

Design by Laura Hough

For Tammy, Noah, and Jeremy

If a man can write a better book, preach a better sermon, or make a better mousetrap than his neighbor, though he builds his house in the woods, the world will make a beaten path to his door.

Ralph Waldo Emerson *(1871 lecture)*

Contents

Notes on the Text and Drawings	xii
Acknowledgments	xiii
Introduction	xv
One. *Gentleman, Choose Your Weapons*	2

Mousetrap *4*, Rat Bell *6*, Tapeworm Trap *8*, Bedbug Killer *10*, Fly Gun *12*, Extendible Fly Swatter *14*, Insect Zapper *16*, Six-Shooter *18*, Plow Gun *20*, Helmet Gun *22*

Two. *Keeping It Clean*	24

Preventive Toilet Seat *26*, Carpet Sweeper *28*, Preventive Siphon Spout *30*, Dog Shocker *32*, Vacuum Cleaner *34*, Washing Machine *36*, Grapefruit Shield *38*, Bedwetting Alarm *40*, Wind Bag *42*, Bird Diaper *44*, Eyeglass Wiper *46*, Religious Soap *48*

Three. *Rube Goldberg Is Alive and Well*	50

Crane Excavator *52*, Elevator *54*, Typewriter *56*, Linotype Machine *58*, Coffin with Escape Hatch *60*, Alarm Bed (I) *62*, Alarm Bed (II) *64*, Rocking Chair Fan *66*, Rocking Chair Butter Churn *68*, Copying Machine *70*, Polaroid Camera *72*

Four. *Necessity, the Mother*	74

Sewing Machine *76*, Umbrella Skirt *78*, Suspenders *80*, Foot Warmer *82*, Luminous Hat *84*, Raincoat with Drain *86*, Clothes Brush Flask *88*, Zipper *90*, Saluting Hat *92*

Five. *Ladies' Liberation* — 94

> Dimple Maker *96*, Bust Supporter *98*, Nose Shaper *100*, Modest Nurser *102*, Responsive Brassiere *104*, Massage Apparatus *106*, Gentleman's Cigarette Holder *108*, Lip Shaper *110*, Centrifugal Delivery Table *112*, Breast Developer *114*

Six. *Way Down (and Way Out) on the Farm* — 116

> Cotton Gin *118*, Combine Harvester *120*, Barbed Wire *122*, Cow Tail Holder *124*, Milking Stool *126*, Hen Goggles *128*, Hen Exerciser *130*, Egg Marker *132*, Sheepdog Nose Hook *134*, Hen Disinfector *136*

Seven. *What Hath God Wrought!* — 138

> Telegraph Code *140*, Telephone *142*, Phonograph *144*, Light Bulb *146*, Electric Burglar Alarm *148*, Electromagnetic Motor *150*, Electric Poison Extractor *152*, Portable Radio *154*, Color Television *156*, Illuminated Brooch *158*, Calculator *160*

Eight. *Loco Motion* — 162

> Air Brakes *164*, Steam Railroad Cattle Prod *166*, Railroad Hammock *168*, Crashproof Railroad Cars *170*, Carburetor *172*, Automobile Chassis *174*, Horse Accessory for Cars *176*, Improved Bicycle Seat *178*, Parking Meter *180*, Horse Taillight *182*, Adjustable Skateboard *184*, Buoyant Boat Chambers *186*, Water Walkers *188*, Submarine *190*, Balloon Birds *192*, Dirigible *194*, Airplane *196*

Nine. *Eureka!* — 198

> Revolving Dinner Table *200*, Scrapbook *202*, Short-Range Parachute *204*, Escape Suspenders *206*, Chewing-Gum Locket *208*, Body Preservation and Display *210*, Improved Paper Clip *212*, Improved Frisbee *214*, Lipstick Dispenser *216*, Muffling Cup *218*

Notes on the Text and Drawings

Patent papers all follow a similar formula: In words and pictures the inventor tells why his device is new and useful, gives a detailed description of how it works, and makes specific claims summarizing its virtues. Language and length, however, differ greatly. Some explanations are quaint, some formal, some pedantic, some abstruse, and many run on for page after page. Thus although all the patent descriptions in this book have been excerpted from the inventors' own words, they have been edited for brevity and clarity. The invention titles are descriptive, and were not always the inventors' own. Patents registered prior to 1836 do not carry a number.

In the case of a well-known invention, the drawing will not always be the very first manifestation but an improved version of what the same or another inventor has already produced. Likewise, while some of the drawings will be easily recognizable, others are strictly diagrammatic or simply amateurish, or have been excerpted from multiple illustrations (which is why not every reference number or letter is explained in the text). But in all cases the drawings shown were the best available reproductions from old documents.

Acknowledgements

The author extends special appreciation to James Poulos III, Esq., and his assistants, Charles Cross and Teresa McKenna, for their research on particular patents; to Michael Congdon and James Raimes for their valuable comments on the manuscript; to Ella Agambar and Louise Pearce for their expert typing; and to Oscar Mastin and the many helpful people at the United States Patent Office for their technical support.

Introduction

Ever since the Massachusetts Bay Colony granted Samuel Winslow the first patent (in 1641, for a novel method of making salt), the American way has been to reward new ideas with exclusive profits. It is not just making a better mousetrap, but making money in the process, which has long been one of our more popular preoccupations. Though it may be true that many patented inventions are more zany than ingenious, and that for every brilliant innovator living handsomely off his royalties there are three dozen would-be Edisons tinkering away in the perpetual poverty of their basement workshops, the sheer multitude of these dreamers and their ideas cannot be easily dismissed.

Through the years Americans have invented at a stupendous clip: almost 5 million U.S. patents have been issued to date, and in one recent year over 100,000 patents were applied for and close to 60,000 granted. (In that year Californians were the most bent on getting rich quick, winning 4,841 patents; Alaskans, perhaps feeling they already had enough pennies from heaven and earth, received the fewest, ten.)

But profit itself is not the only stimulus. Inventors get a certain visceral excitement from solving problems: Whether creating gadgets, designing machinery, or developing processes, there is genuine intellectual joy in being the first to come up with something that can be perceived as useful. Indeed every device awarded a patent must demonstrate its utility as well as its novelty. "The world is not interested in a bright idea," an American inventor once said, "just in whether it works."

What separates inventors from the rest of us is their gift not only for visualizing what's new, but for acting upon the inspiration. Theirs is an amalgam of energy, ego, perspicacity, and luck. "I don't much like to work," said one inventor whose patents earned and lost him millions. "I'd rather think and daydream. I left school over forty years ago and I've never had a job. That's my greatest claim to fame." The rest of us are more myopic: We often can't see the *problem,* much less the solution. For those who do get beyond wanting to leave well enough alone, the kibosh comes when they see an invention that appears so simple and obvious that they are left muttering, "Why didn't *I* think of that?"

Obviously, though, while some inventions catch on, like Mark Twain's suspenders (1871, see p. 80), some don't, like Bertha Dlugi's bird diaper (1959, p. 44). While some make fortunes, like Thomas Edison's speaking machine (1878, p. 144), some won't, like Alfred Clark's combination rocking chair/butter churn (1913, p. 68). And while some have revolutionized America, like Henry Ford's automobile chassis (1901, p. 174), some haven't, like Martin Goetze's dimple maker (1896, p. 96). As the reader will quickly see, however, the line between utility and zaniness is often very fine indeed.

Why have patents? A country without good patent laws, wrote Twain, "is just a crab and can't travel anyway but sideways and backways." In truth patents are necessary because of competition and profit. Many if not most inventions are derivative in nature, and the "improved" versions are usually cheaper to produce than the originals, although the subsequent developer hasn't had to invest as much time, effort, and money in research and prototype. In short, if there were no protection offered by law, inventors would be much less motivated.

The incentive factor was recognized as far back as 1474, by the governors of Venice:

> There are in this city, and also there come temporarily by reason of its greatness and goodness, men from different places and most clever minds, capable of devising and inventing all manner of ingenious contrivances. And should it be provided, that the works and contrivances invented by them, others having seen them could not make them and take their honour, men of such kind would exert their minds, invent and make things which will be passed that . . . each person who will make in this city any new and ingenious contrivance, not made heretofore in our dominion, as soon as it is reduced to perfection, so that it can be used and exercised, shall give notice of the same. . . . It being forbidden to any other in any territory and place of ours to make any other contrivance in the form and resemblance thereof, without the consent and license of the author up to ten years.

The American system was created in September 1787, when the Constitutional Convention adopted Article I, Section 8: "Congress shall have power . . . to promote the progress of science and useful arts by securing for limited times to authors and inventors the exclusive right to their respective writings and discoveries." Thomas Jefferson, an early champion of the idea that "ingenuity should receive liberal encouragement," became the first patent examiner. Perhaps out of a sense of conflicting interests, Jefferson declined to take out patents on his own inventions (which included the swivel chair and the pedometer).

In August 1814, as the British were burning Washington, Dr. William Thornton (then the superintendent of patents) saved many early drawings and models by persuading the British commander not to burn "what would be useful to all mankind." Patents proliferated in the early part of the nineteenth century, to the point where a patent official is said to have resigned with the exclamation that "everything has been invented!" (The Patent Office claims this story is apocryphal.)

Most modern countries have their own patent systems. Switzerland and Japan modeled theirs on America's. "We have looked about us to see what nations are the greatest," remarked a Japanese commissioner in 1900, "so that we can be like them. We said, 'What is it that makes the United States such a great nation?' and we investigated and found that it was patents." Indeed, among non-Americans holding U.S. patents, the Japanese are far ahead of anyone else: In 1983 they were granted 8,632 of them.

Even the Russians, who claim to have invented everything from teapots to television, found that without certain capitalistic incentives Ivan the Inventor would have little inclination to remember his bright new ideas. Nowadays Soviet inventors are awarded up to 25,000 rubles (about $40,000) for each patent; one Russian is reputed to have earned over a million rubles to date.

How does the process work? If a device is deemed new and useful, its inventors are given the right to exclude all others from manufacturing the product for seventeen years. That figure was arrived at in 1861 as a compromise between those who wanted to retain the original fourteen-year period with a (commonly granted) seven-year extension, and those who wanted to abolish the extensions altogether. The right conferred by the patent grant extends throughout the United States and its territories and possessions.

In the language of the statute, "any new and useful process, machine, manufacture, or composition of matter, or any new and useful improvements thereof" can be patented. New varieties of plants are also eligible, as are designs. The Atomic Energy Act of 1954 excludes the patenting of inventions useful solely in atomic weapons. Devices that are not considered useful, such as perpetual-motion machines, are refused patents. So are methods of doing business, and mixtures of ingredients. (Thus "patent medicines" are not patented medicines.)

An applicant will also be denied a patent if his invention has been described in a printed publication anywhere in the world, or if it has been in public use or on sale in this country either prior to when he produced his invention or more than one year before he filed his application.

That is one reason why the Patent and Trademark Office maintains a cavernous library (in Arlington, Virginia) with over 120,000 volumes of technical books and 90,000 bound volumes of periodicals. There is also a large search room where the public can examine all U.S. patents granted since 1836, when the consecutive numbering system was inaugurated. Now patents are arranged according to some 300 classes and 112,000 subclasses, enabling inventors to determine whether somebody else has already come up with their idea. Those who cannot make it to the search room can inspect copies of the patents at one of the Patent Depository Libraries, located in thirty-three states across the country.

Although inventors may prepare and file their own applications for patents, most employ the services of patent attorneys or agents to help them through the quagmire of regulations and possible litigation. (About two-thirds of all patents applied for are granted; about one percent of these end up being challenged in court.) According to the law, only the inventor or his legal representative may apply for a patent. If an inventor is insane, his guardian may apply. Should two or more people have made the invention, they can apply as joint inventors; someone who has made nothing more than a financial contribution to the effort is not considered a joint inventor. To further ensure the integrity of the process, employees of the Patent Office are prohibited by law from applying for or acquiring (except by inheritance) any patent.

Boiled down to its barest essentials, the application requires a written document, a drawing, and the filing fee. The written document contains a detailed description of the invention together with an oath or declaration that it is new and useful. The drawing must ordinarily be in pen and ink (black and white); models were required until about 1880, when it was decided that they were too expensive to make and difficult to store. The basic filing fee in 1985 was $150 for small entities, $300 for others.

Currently it takes about twenty-five months from the time a patent is applied for until the time it is granted (for an issuance fee of $500). The terms *Patent Applied For* or *Patent Pending* have no legal effect, unless they are used falsely, but the person to whom a patent is awarded must mark his invention with that word if he subsequently wants to recover damages from an infringer.

The competition is often keen. When two or more applications are filed by different people who claim substantially the same invention, the government initiates an "interference" proceeding to determine which inventor came first. Such proceedings can be overwhelmingly tedious. For example, when the Patent Office had to determine the original inventor of the telephone, it spent two years and heard hundreds of thousands of words of testimony. In the end Alexander Graham Bell's claims of priority were fully substantiated.

Avoiding such litigation is perhaps one reason why many inventors assign their rights to corporate manufacturers, whose legal staffs are highly mobile and specialized and whose research and development departments are by now the largest group of patent awardees. Another is that it takes so long and costs so much to make and market a product successfully. The seventeen-year patent period is seldom enough time for the lone inventor, toiling away in his garage or laboratory, to achieve financial security.

Not to mention the pitfalls of unbridled optimism. One American inventor tells the story of how he came to the Patent Office in 1938 with a three-dimensional motion-picture system. The examiner yawned, opened a long file drawer, casually pulled out a folder, and asked, "Is this what you mean?" It was exactly the same as the applicant's invention, almost word for word. And it had been issued in 1919!

It is not only the priority of time that separates successful inventions from those that attract little more than dust in Patent Office files. With the benefit of hindsight, it is easy to admire (and envy) how a winning inventor shrewdly anticipated a waiting market and ingeniously created the means by which it could be tapped or exploited. But not everyone has an easily stretchable imagination; many people would initially view as needless or farfetched what inventors and the government may deem new and useful. The public can dismiss the newfangled as eccentric; the Patent Office must credit its potential.

This book celebrates both the genius and the whimsy of American inventiveness. It is a sampler of brilliant and bizarre inspirations, the brainchildren of dedicated tinkerers—some of whom have gained, some of whom have lost, but all of whom have had enough gumption to venture into the world with their patented problem solvers.

The world, alas, has not always been respectful. Contrasted here are some classic devices (shown in their original or "improved" versions) that have been great commercial successes—indeed may have changed the way we live—with some that haven't. From mousetraps to muffling cups, however, they were all granted U.S. patents. Whether the less fortunate inventors presented here should have become rich, or the unknowns famous, is left entirely to the reader's judgment.

MOUSETRAPS AND MUFFLING CUPS

ONE
Gentlemen, Choose Your Weapons!

Man sometimes uses his superior reasoning ability to better himself by doing away with his beastly four-legged counterparts and, on occasion, with other members of his own species. Prehistoric evidence of his carnivorous nature traces well back to the vegetarian paradise of Eden. It is uncertain exactly how the snake got his due or what Adam and Eve used for fly swatters, but archaeologists have dated early animal traps to 200,000 B.C., and barbed fishhooks—remarkably like modern ones—existed in Egypt in 3,000 B.C.

One could have made the argument, way back when, that the rock with which Cain slew Abel was sufficient means for dispatching fellow creatures. Enough, after all, was enough. But mankind quickly learned that survival belonged to those with the best weapons (later called

> We have met the enemy and they are ours.
> *Oliver Hazard Perry*
> (announcing victory at the
> Battle of Lake Erie, 1813)

defense systems), and it was not all that long before the rock was replaced by the nuclear missile. In between there were the spear, sword, bow and arrow, gunpowder, and shrapnel shell. Soon after the patent system came into being appeared the revolver (1836), rifle (1848), and torpedo (1866); somewhat later the ballistic missile (1943) and the atomic bomb (1945).

Throughout history, perhaps even more prevalent than the problems of guns and butter has been the question of what to do about bugs and mice. We've never been able to get rid of either, at least not to the point where anyone could say, "Why not leave well enough alone?" After all these years and inventive killing devices, man is still trying to build a better mousetrap.

Mousetrap
Patent No. 661,068 (1900)
Charles F. Nelson of Galesbury, Illinois

It has been customary heretofore to set traps for catching mice, rats, and similar animals by locating the bait in a comparatively unprotected position upon the trigger or base of the trap and so that when the animal nibbles or pulls at the bait a spring-jaw will be freed and fly back to engage and hold the animal between itself and the base. It has been found that in some instances the bait will be removed and eaten by the animal without springing the trap, and it is to provide a trap of simple construction in which the bait cannot be removed without releasing the engaging jaw that my present invention is particularly directed.

The invention also has in view to provide a trap of simple construction which can be manufactured inexpensively in a very few parts and assembled to provide a trap adapted to be easily and quickly set and one which is effective in practical use.

To set the trap, the engaging jaw is carried forward into the position in Fig. 1, and then the locking-arm is swung over the engaging jaw and its end engaged under the lug H of the trigger. The spring tension of the engaging jaw exerted upwardly against the locking-arm is sufficient to keep the arm in its locked engagement with the trigger and the bait-plate of the trigger in the lifted position shown in Fig. 2, so that when the animal endeavors to get at the bait and presses upon the trigger the lug H and locking-arm will be disengaged, permitting the engaging jaw to fly back and strike and hold the animal against the base.

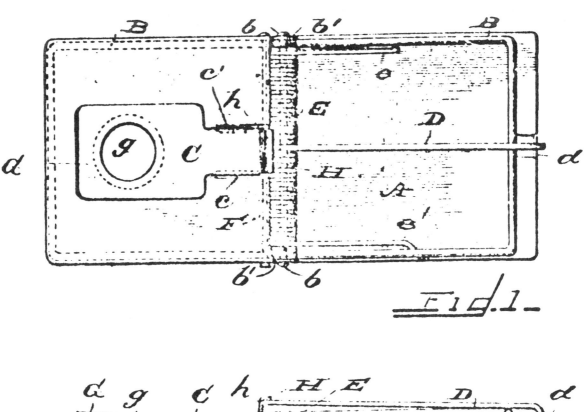

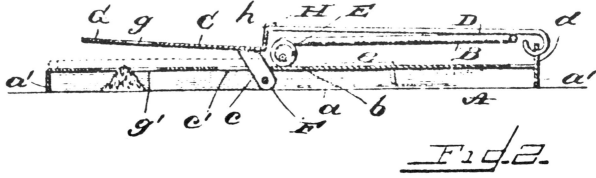

Rat Bell
Patent No. 883,611 (1908)
Joseph Barad and Edward E. Markoff of Providence, Rhode Island

Our invention relates to improved means for exterminating rats, mice, &c., and it consists essentially of a device having a frame, an endless flexible resilient band or collar supported by and encircling part of the frame, and spring-resisted tripping means operatively connected with a member of the frame, all constructed and arranged whereby an animal, say a rat, upon introducing its head through the frame opening and seizing the lure or bait attached to said tripping means automatically releases the latter, which action at the same instant also releases and separates the said frame member and frees the expanded band, which latter then immediately contracts around the animal's neck before he can retreat from the device or apparatus. The thus bedecked animal is not caught or confined in any manner whatever but is free to return to its hole and colony. The "bell-rat," as it may be termed, then in seeking its burrow or colony announces his coming by the sounds emitted by the bells, thereby frightening the other rats and causing them to flee, thus practically exterminating them in a sure and economical manner. It may be added that the spring-band or collar is not liable to become accidentally lost or slip from the rat's neck because the adjacent hairs soon become interwoven with the convolutions of the spring to more firmly hold it in place. [See Figs. 1–5.]

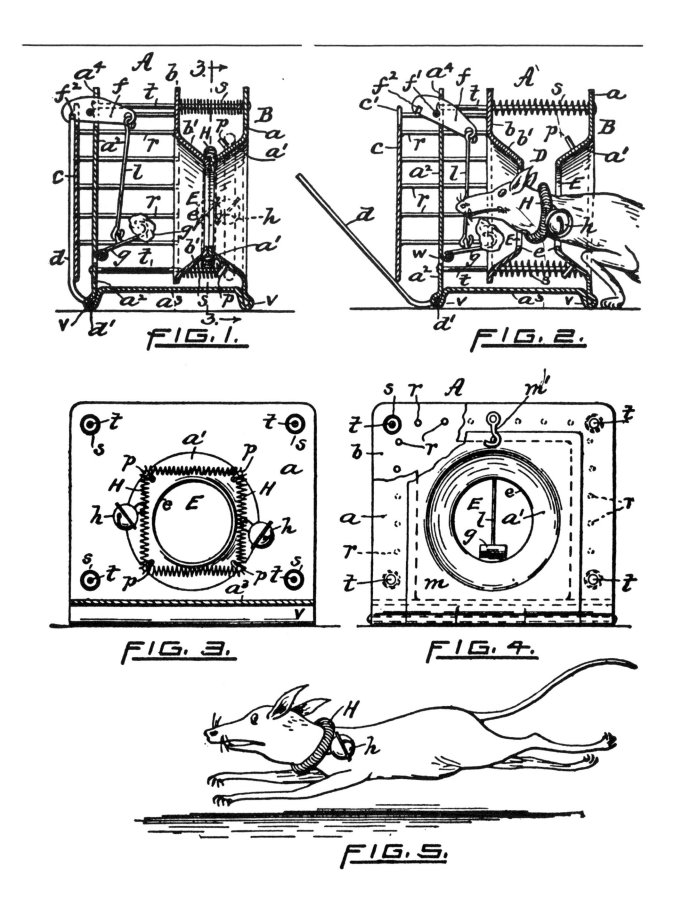

Tapeworm Trap
Patent No. 11,942 (1854)
Alpheus Myers, M.D., of Logansport, Indiana

The object of my invention is to effect the removal of worms from the system, without employing medicines, and thereby causing much injury.

My invention consists in a trap which is baited, attached to a string, and swallowed by the patient after a fast of suitable duration to make the worm hungry. The worm seizes the bait, and its head is caught in the trap, which is then withdrawn from the patient's stomach by the string which has been left hanging from the mouth, dragging after it the whole length of the worm.

The trap is baited by taking off the cover b, of the exterior box, and filling the interior box with the bait which may consist of any nutritious substance. The interior box d, is then pushed down until the stud f, catches between the teeth of the opening e, and holds it with the openings, e and c, opposite each other, the points of the teeth being then below the lower edge of the opening c. The trap, having the cord h, attached to a ring i, on the lid is then swallowed. The worm, in inserting its head at the opening e, and eating the bait, will so far disturb the inner box as to work it free of the stud f, when the box will be forced upward by the spring g, and the worm caught behind the head, between the serrated lower edge of the opening in the interior box, and the upper edge of the opening in the exterior box. The trap and the worm may then be drawn from the stomach, by the cord h. [See Figs. 1 and 2.]

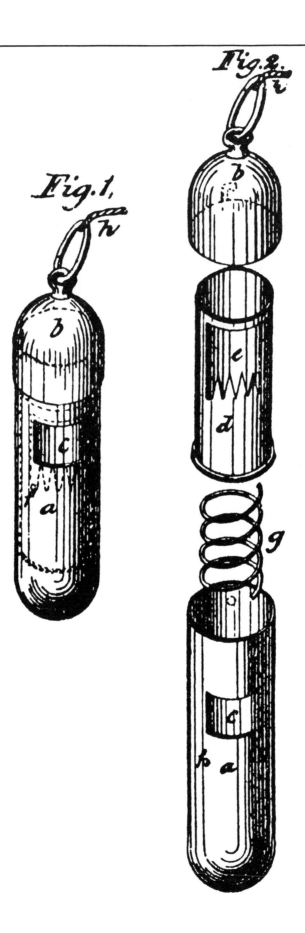

Bedbug Killer
Patent No. 616,049 (1898)
Frank M. Archer of New York, New York

This invention relates to bedbug-exterminators and it consists of electrical devices applied to bedsteads in such a manner that currents of electricity will be sent through the bodies of the bugs, which will either kill them or startle them, so that they will leave the bedstead. The electrical devices used consist of a battery, induction-coil, a switch, and a number of circuits leading to various locations on the bedstead, where are placed suitable circuit-terminals, arranged so that the bugs in moving about will close the circuit through their own bodies.

The space c^2 between the rings is such that a bug in crossing from one to the other must close the circuit through its own body, and thus receive a current of electricity. If these rings are placed on a leg of the bedstead, an insect in climbing up will when it receives the shock more than likely change its mind and return in the direction whence it came. Another location where the contacts would be particularly efficient is at the joints between the side pieces and the head and foot boards. A perspective of such a joint is shown in Fig. 3. A pair of insulated contact-strips e and e^1 is placed along each of the contiguous edges, so that the insect in crossing a pair or the adjacent members of the two pairs will necessarily receive a current, which will either terminate its career at once or make it seek other locations.

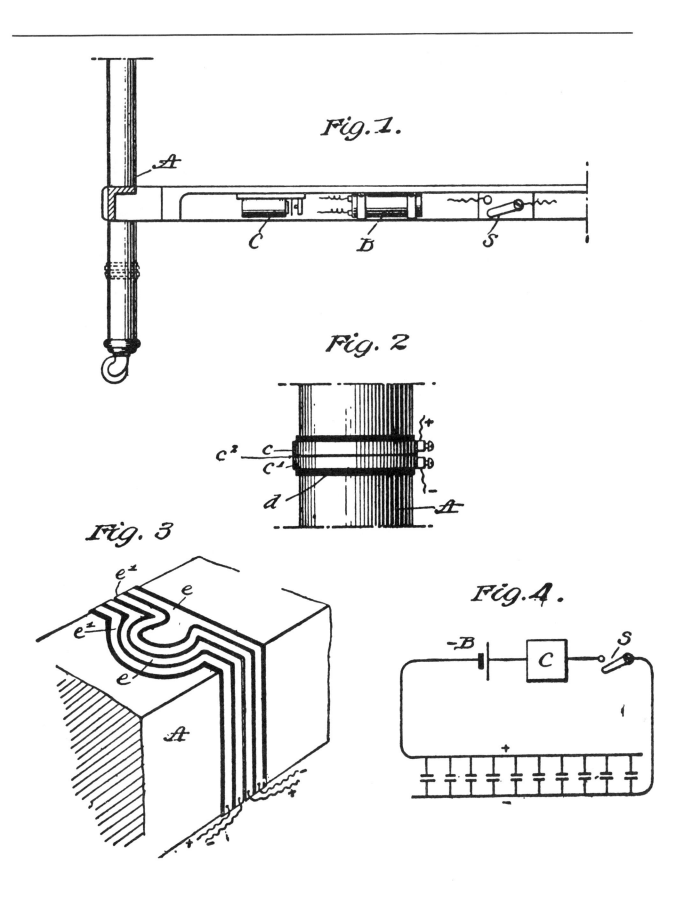

Fly Gun
Patent No. 1,468,373 (1923)
George W. Blake of Wyandotte, Michigan

This invention aims to provide a fly swatting device constructed somewhat on the principle of a toy which will afford considerable amusement to juveniles and at the same time contribute to the extermination of flies and similar insects. The device is made to represent a fire arm having a projectile provided with a swatting or imprisoning member which is articulated relative to the projecting part of the device so that said swatting member may more or less accommodate itself to surfaces of various contours in order to swat, imprison or kill the fly or insect toward which said member is projected.

My invention further aims to provide a fly swatting device calculated not to injure the surface against which it impinges and the swatting member of the device is adapted to be forcibly projected, when released, with such impetus as to preclude the escape of any insect at which the swatting member is aimed. To forcibly actuate the swatting member of the device a spring is employed and embodied in the frame or body of a fire arm, particularly a pistol, having a trigger mechanism which controls the action of the spring. [See Figs. 1–3.]

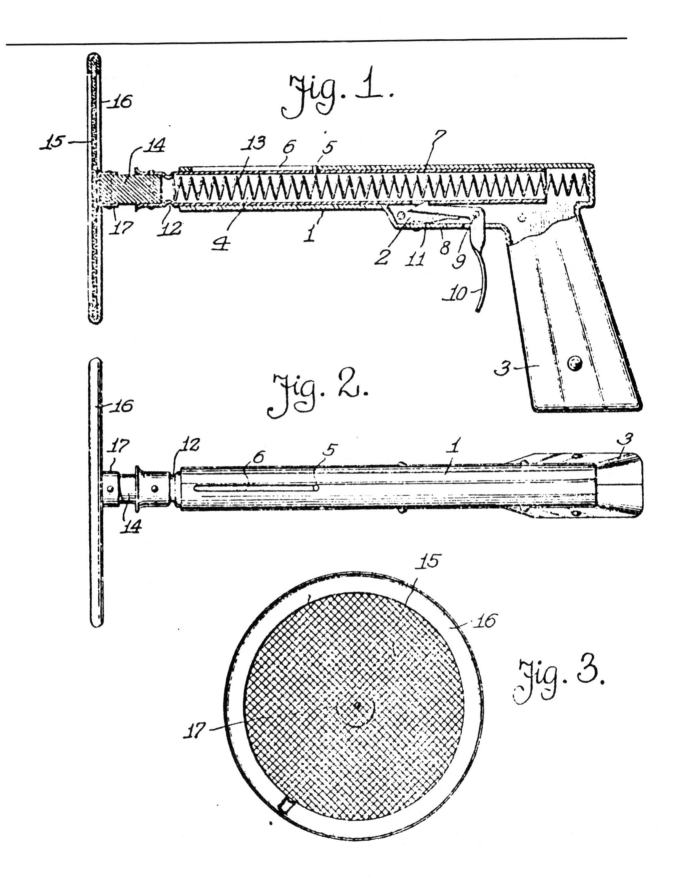

Extendible Fly Swatter
Patent No. 685,990 (1928)
Walter H. A. Emanuel of Fairchild, Wisconsin

This invention relates to new and useful improvements in fly swatters, and has for its primary object to provide a device of this character wherein the handle portion thereof is readily extendible so that flies or other insects at high inaccessible points may be reached by merely extending the handle.

A further object is to provide a swatter of this character that is extremely simple of construction, inexpensive of manufacture, and highly novel and useful for the purposes intended.

As clearly disclosed in both Figures 1 and 2, I arrange a ring 9 over one of the rods 6 after which the ends of the rods outwardly of the ring are inserted within the tubular section 7 of the handle. The shank and tubular portions are then adjusted with respect to each other so that the desired length of the handle may be secured after which the ring 9 is moved toward the adjacent end of the tube 7 for causing the rod 6--6 to be forced apart directly at the end of the tube 7 which will of course wedge said rods at this point within the tube for preventing the movement of the sections with respect to each other.

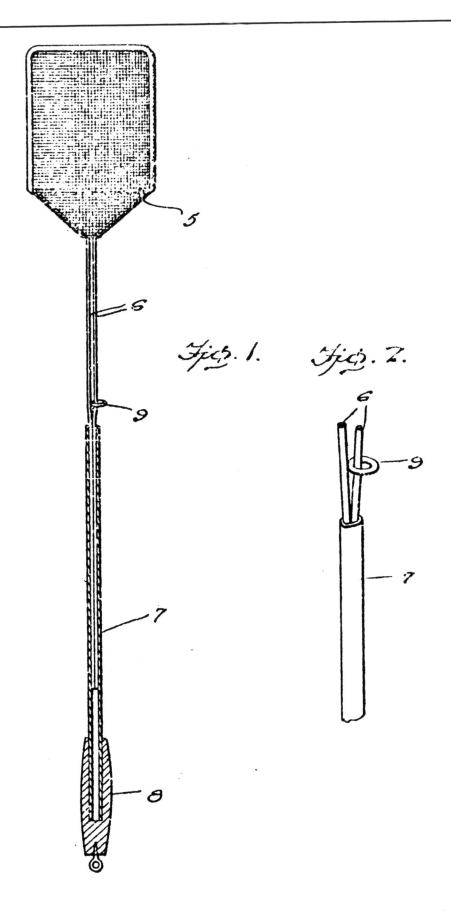

Fig. 1. Fig. 2.

Insect Zapper
Patent No. 2,835,071 (1958)
Donald E. Partridge of Los Angeles, California

It is an object of the present invention to provide an improved electrical insect destroyer which is of simple construction and hence relatively inexpensive to manufacture.

It is a principal object to provide an insect electrocutor having a pair of substantially upright grids through which an insect in flight may pass, said grids having a potential difference applied thereto of a value insufficient to cause a spark to jump the gap between the grids until an insect passing between the grids shortens the potential path and thereby electrocutes the insect.

The insect electrocutor of the invention has utility in a variety of places where insects may be found and is not unattractive in its ornamental aspects as well as strictly utilitarian. It may be placed in any desired location and is not obtrusive. Thus its uses may be itemized at grocery stores, on the meat and vegetable counters, cutting rooms, food processing rooms, or wherever flies and other like insects are wont to gather. Likewise, it may be hung from the ceiling, for example in a dairy barn.

The particularly novel feature of the generally upright grids makes the device, or one patterned after the lines of the upright grids, exceedingly efficacious in exterminating insects in flight when attracted to feeding grounds such as chicken coops and the like. [See Figs. 1–3.]

It will ordinarily not be necessary to bait the insect destroyer. As soon as a few insects or flies have been electrocuted, the tray and the grids will be covered with the vomit and excreta of the insects which will attract more insects. The tray should be periodically cleaned and the destroyed insects disposed of.

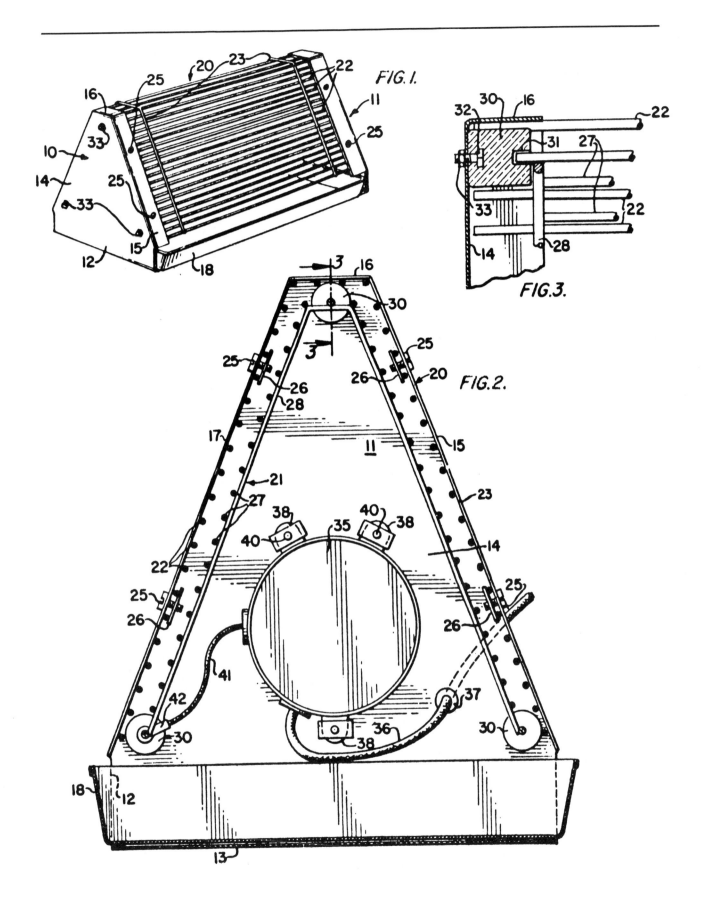

Six-Shooter
Patented in 1836
Samuel Colt of Hartford, Connecticut

Among the many advantages in the use of these guns, independent of the number of charges they contain, are, first, the facility in loading them; secondly, the outward security against dampness; thirdly, security of the lock against the smoke of the powder; fourthly, the use of the partitions between the caps, which prevent fire communicating from the exploding cap to the adjoining ones; fifthly, by the hammer's striking the cap at the end of the cylinder no jar is occasioned, deviating from the line of sight; sixthly, the weight and location of the cylinder, which give steadiness to the hand; seventhly, the great rapidity in the succession of discharges, which is effected merely by drawing back the hammer and pulling the trigger.

Division 1 of the drawings represents a pistol. . . . Division 5 represents the mechanical combination of the entire instrument.

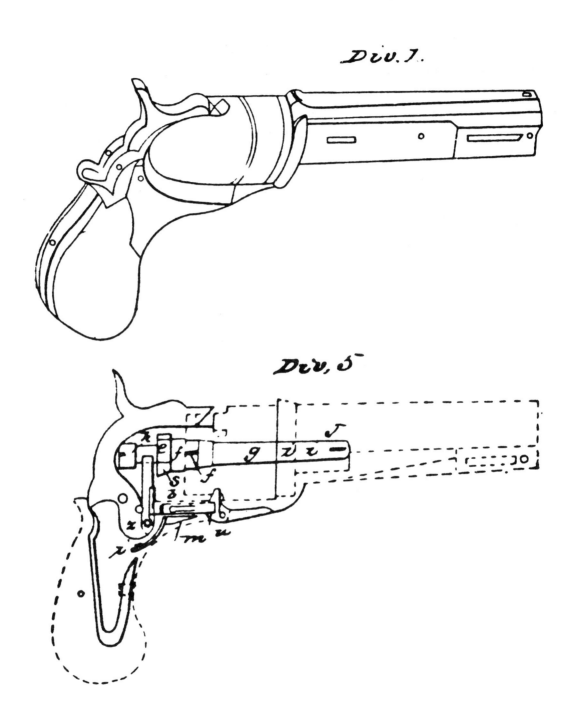

Plow Gun
Patent No. 35,600 (1862)
C. M. French and W. H. Fancher of Waterloo, New York

The object of our invention is to produce a plow equal, if not superior, in point of strength and lightness to that implement as ordinarily made, and at the same time to combine in its construction the elements of light ordnance, so that when the occasion offers it may do valuable service in the capacity of both implements.

Its utility as an implement of the twofold capacity described is unquestionable, especially when used in border localities, subject to savage feuds and guerrilla warfare. As a means of defense in repelling surprises and skirmishing attacks on those engaged in a peaceful avocation it is unrivaled, as it can be immediately brought into action by disengaging the team, and in times of danger may be used in the field, ready charged with its deadly missiles of ball or grape. The share serves to anchor it firmly in the ground and enables it to resist the recoil, while the hand-levers *A* [see Figs. 1 and 2] furnish convenient means of giving it the proper direction.

Fig. 1.

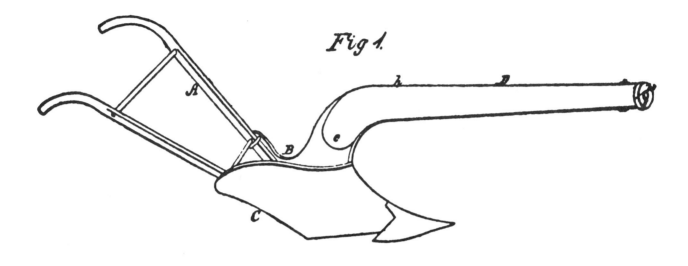

Fig. 2.

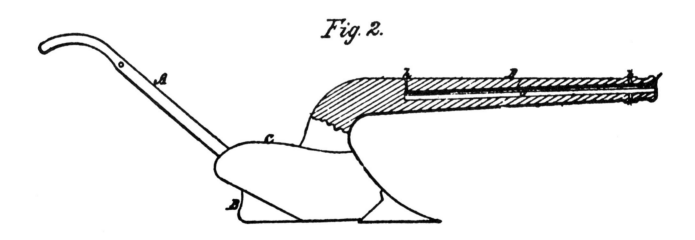

Helmet Gun
Patent No. 1,183,492 (1916)
Albert B. Pratt of Lyndon, Vermont

This invention relates to weapons, and among other objects provides a gun adapted to be mounted on and fired from the head of the marksman.

Figure 1 is a side elevation of a soldier's helmet equipped with a gun embodying the invention.... Figs. 11 and 12 show different devices for operating the gun trigger from the mouth of the marksman.

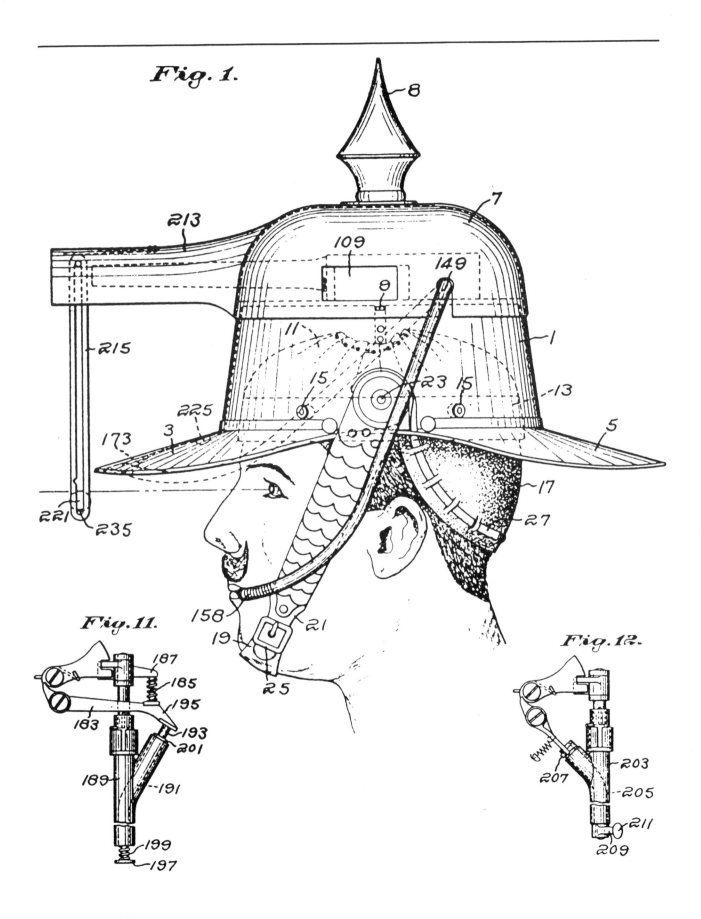

TWO
Keeping It Clean

Human beings have always prided themselves on having certain enduring (if not endearing) qualities by which they can be distinguished from other kinds of animals. Arguable though such distinctions may be, throughout history they have gained widespread acceptance: One is the ability to reason, another the desire to keep things clean and orderly.

Small wonder, then, that some of mankind's most popular inventions have had to do with soap and water. As early as 3000 B.C., or so we're told, the Sumerians boiled up a concoction with which they could lather themselves and scrape off the grime. The Phoenicians improved upon the formula, the Romans developed their baths, and by the Middle Ages soap was in general use—though it wasn't until the seventeenth century that the first soap patent was issued in England. Indeed bathing in England was infrequent before the Baths and Washhouses Act of 1846; prior to that, the rich used heavy perfume, and the poor simply stank.

Chamber pots were in use throughout Europe until late in the eighteenth century, when the first water closet was made. But the problem of disposal became steadily worse until sanitary engineering was developed during the Victorian age. Invisible organisms were being

> We are charmed by neatness.
> *Ovid*, Ars Amatoria

examined by the 1860s, and the germ theory of disease was gaining credence among reputable scientists—among them Louis Pasteur and Joseph Lister. Seeking a way to retard the abnormal fermentation of wine and beer, Pasteur applied his ideas to the treatment of milk and came up with the process still used today. Lister developed a carbolic spray to kill all germs in the surrounding air; by 1890 he had narrowed the need for destroying microbes to the area of infection only.

Meanwhile M. R. Bissell, owner of a china shop in Grand Rapids, Michigan, was seeking a cure for the allergic headaches he suffered from the dusty straw in which his new crockery was packed. In 1879 he invented a sweeper with adjustable brushes, attached to a box that would collect and hold the dust (see p. 28). Bissell stopped sneezing and became rich. Not long after came the vacuum cleaner (in 1901, p. 34), the motorized washing machine (in 1914, p. 36), and the spin dryer (1924).

No doubt the inventors of the other items pictured in this chapter had an equal abhorrence of dust and dirt. Apparently, however, the only people who agreed that their cleaning devices had any possible merit were the government examiners who awarded them patents.

Preventive Toilet Seat
Patent No. 90,298 (1869)
Francis Peters and George Clem of Cincinnati, Ohio

This invention relates to a device which renders it impossible for the user to stand upon the privy-seat; and consists in the provision of rollers on the top of the seat, which, although affording a secure and convenient seat, yet, in the event of an attempt to stand upon them, will revolve, and precipitate the user on to the floor.

[In Figs. 1 and 2] *A* represents the box, having one or more pairs of standards, *B*, which afford journal-bearing for a roller, *C*, over the front-edge of the box, and, where necessary, of side rollers *C' C'* and a back roller, *C"*.

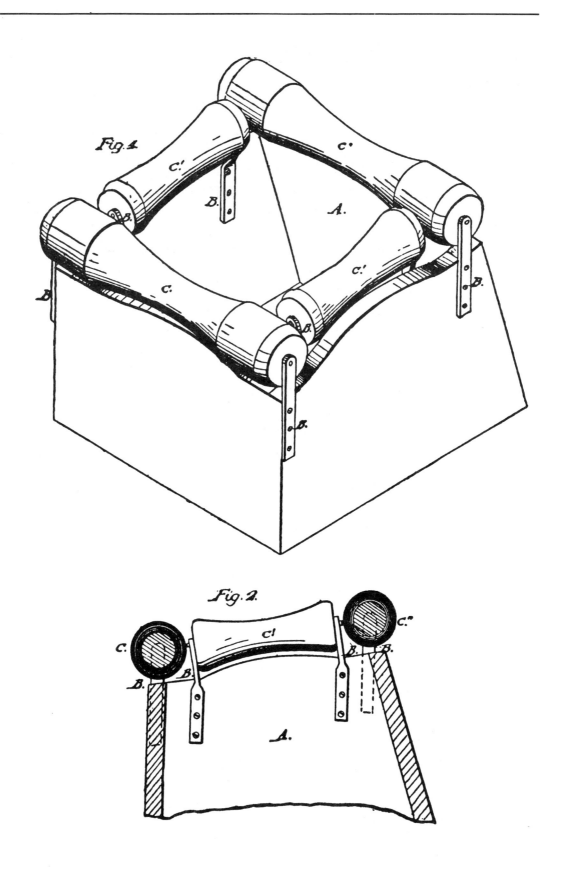

Carpet Sweeper
Patent No. 217,854 (1879)
Melville R. Bissell of Grand Rapids, Michigan

The nature of my invention relates to a carpet-sweeper having the dust-pans attached together in some suitable manner, and so constructed that they can readily be detached from the frame or case of the sweeper.

The object of my invention is to enable the operator to readily detach and remove the dust-pans from the sweeper, in order to empty the contents of the pans.

Figure 1 represents a perspective view of my invention complete and ready for use. Fig. 2 represents a perspective view of the two dust-pans attached to the cover of the sweeper and detached from the sweeper case or frame.

In Fig. 1, B represents the frame or case of the machine; H, the handle, and A the brush-roller.

In Fig. 2, $P\ P$ represent the dust-pans, the outer sides of the pans being directly attached to the case-cover D, and the inner sides of the pans being attached to the cover by means of the braces $W\ W$. The cover D, when constructed as above described, may be provided with a handle, C, in order more readily to lift it, with the pans, from the case. It may also be provided with a catch at either end, for the purpose of attaching it more firmly to the case.

This device saves the trouble of carrying the entire sweeper, and enables the user to empty the dust readily into the stove or any small receptacle without spilling or litter.

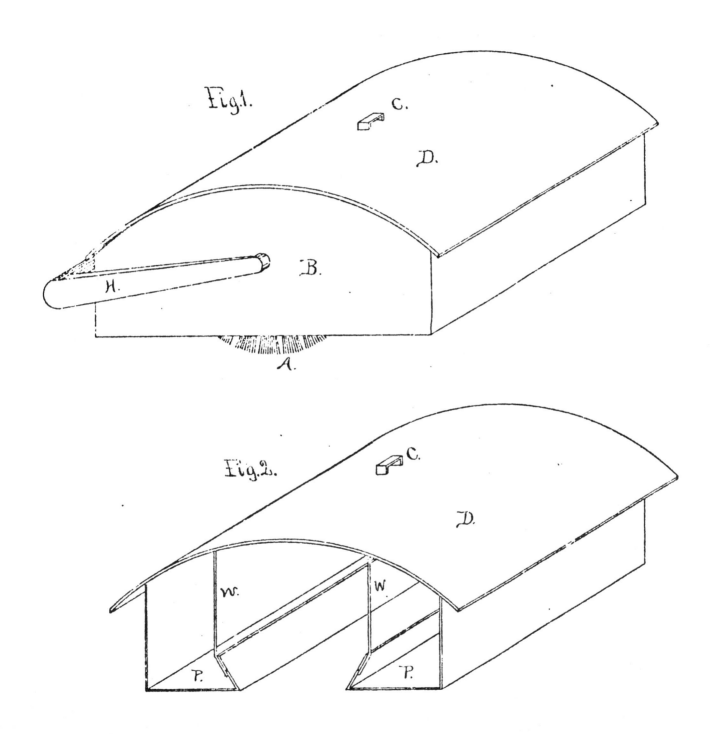

Preventive Siphon Spout
Patent No. 641,201 (1900)
Josef Fleischer of New York, New York

In using the siphon before my invention nothing prevented the insertion of the siphon-mouth in any opening of the body to be used as a syringe, thereby contaminating it with germs of disease. With my improvement in using the siphon the projections *a* [see Fig. 1] in no way interfere; but if it is attempted to misuse the siphon—e.g., as a syringe—the projections *a* prevent an insertion in or application to any part of the body.

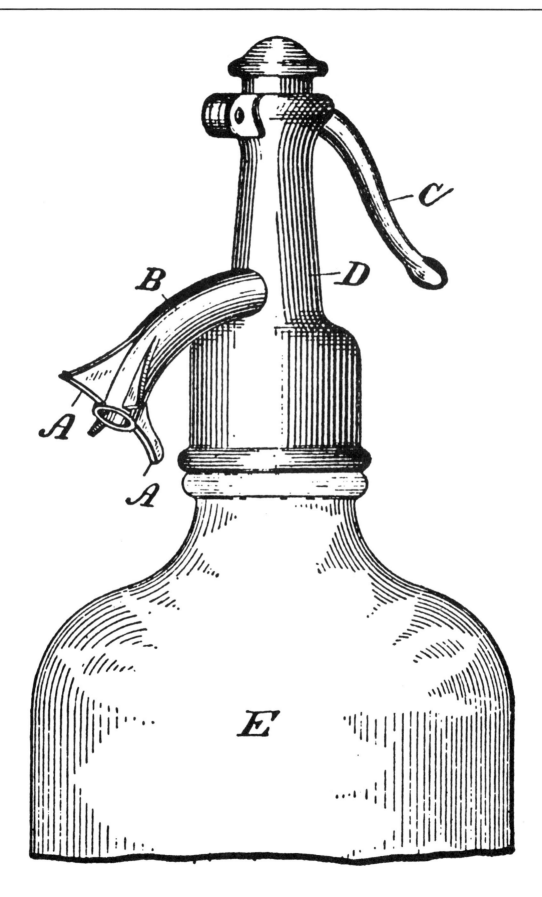

Dog Shocker
Patent No. 922,956 (1909)
Aden H. Roberts of Waverly, New York

This invention relates to improvements in devices for preventing dogs from committing the nuisance of urinating against buildings, walls, and other structures; and my object is to provide a device which may be conveniently placed at points where such nuisance has been, or is likely to be, committed, whereby an electric shock will be administered to a dog, when attempting to commit such nuisance, which will effectually prevent a recurrence of the act in that locality by the dogs so punished.

The plate [Fig. 1], when coupled to an electric light circuit, or to a battery or other source of electric current of suitable power, becomes a terminal from which no current will pass until a ground connection is made. When, therefore, the device has been placed in position in front of a building, or the like, where dogs have been in the habit of committing nuisance, or are likely to commit nuisance, the next dog that attempts the act, will receive a severe shock the instant the stream of urine strikes the plate, by reason of the grounding of the current through the dog's body. After receiving one such shock it is believed that that particular locality will be shunned in the future by every dog so punished.

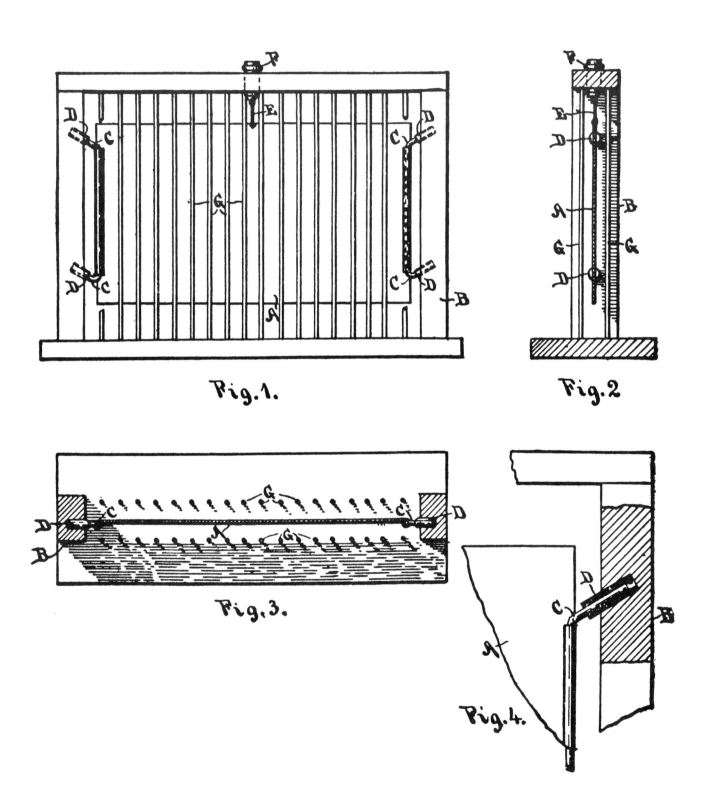

Vacuum Cleaner
Patent No. 1,209,722 (1916)
James B. Kirby of Cleveland, Ohio

Experience has shown that a sweeping or brushing device located within the inlet mouth of the apparatus and operated by the air pressure can be operated efficiently only in case it be very delicately balanced or else be supplied with a very powerful turbine; in case the brushes shall rest upon the floor with slightly too much force, it will be found impossible to operate them except with a turbine of such size as to detract seriously from the efficiency of the cleaner, while if they be suspended so as to engage the floor with the smallest force at which they will be effective in picking up strings, lint, etc., a very small accident to the adjustment will render them ineffective. Also in case the brushing device should be so nicely balanced within the shoe as to operate satisfactorily and also develop a proper brushing effect, the shoe must necessarily be kept in constant motion since a stoppage of the shoe at any point for more than a very brief interval will result in such a wearing effect upon the nap of the floor covering as to give rise to a noticeable mark.

The objects of the present invention are, therefore, the provision of a new and improved construction of brushing mechanism; the provision of new and simplified means for mounting the same within the shoe; the provision of means for preventing the operation of the brushing mechanism while the shoe is stationary and permitting the same while the shoe is moving; the provision of a new and improved construction of air conduit, whereby the motor duct can be conveniently closed in case the brushing device is not desired.

Fig. 3 is a vertical cross-sectional view taken through the center of the head upon the line 3—3 of Fig. 1 and looking in the direction of the arrows; Fig. 4 illustrates the same head reversed for the attachment of a hose connection, the main part being shown in cross section and the air slots being closed.

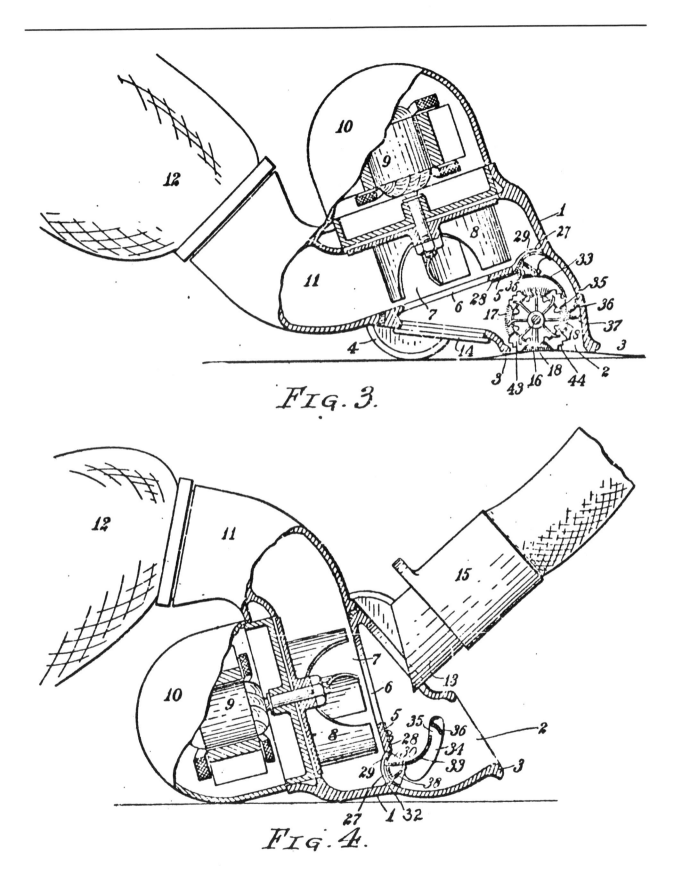

FIG. 3.

FIG. 4.

Washing Machine
Patent No. 1,318,981 (1919)
Robert W. Donley of Philadelphia, Pennsylvania

The invention comprehends a washing machine consisting of few simple parts which cannot readily get out of order, the machine being so constructed that it can be readily operated with a minimum of manual or other labor and in which furthermore, the clothes or fabric, will be subjected to the action of the cleansing fluid so that the latter will be forced through the clothes or the fabric thereof and thus thoroughly cleanse the same.

In the use of the device described the outer receptacle is suitably filled with water or some other cleaning fluid and the clothes holding member 27 having been removed from the inner receptacle by raising the handle, rotation of the operating member will cause the holder to be screwed down thereon . . . thus securely gripping and holding the clothes against the underside of the latticed frame, as shown in Fig. 1. A suitable cover 42 is provided to close the outer receptacle and also cover the inner receptacle, and the stem 26 passes through the cover, as will be clearly seen in Fig. 1. The operator now actuates the handle 15 and on account of the clothes contained in the inner receptacle, an upward movement of the handle will create a suction which will raise the free portions of the flap valve, so that the water can pass through the apertures 23 into the inner receptacles and then when the handle is subsequently moved down again, the water received in the inner receptacle being unable to return to the outer receptacle through the apertures 23 on account of the closing of the flap valve thereon, will be forced up through the clothes, and should the inner receptacle become filled with water by this action, the water will flow over the edge of the inner receptacle back into the outer receptacle, indicated by a series of arrows in Fig. 1.

Therefore continued reciprocation of the handle will always bring in a new or fresh quantity of water to the inner receptacle from the outer receptacle.

The outer receptacle will of course, be drained through the pipe 37 when the spigot 38 is opened, and in case it is desired to rinse the clothes by filling the outer receptacle with fresh water, the spigot 40 can be opened and the water drawn into the inner receptacle and through the clothes, will be ejected through the pipe 39.

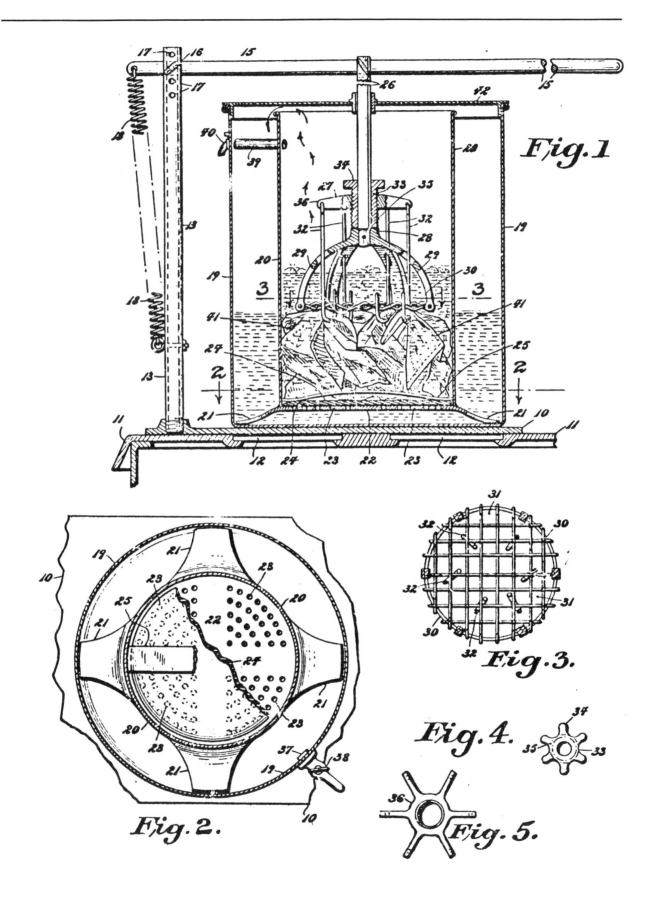

Grapefruit Shield
Patent No. 1,661,036 (1928)
Joseph Fallek of Brooklyn, New York

This invention relates to dining accessories, and has particular reference to a shield for attachment to a halved grapefruit or other similar citrus fruit for preventing spattering of the juice when eating.

The invention furthermore comprehends a grapefruit or similar citrus fruit shield which is extremely simple in its construction, inexpensive to manufacture and which is highly efficient in its purpose.

In practice, the lower end 11 of the shield 10 is arranged in embracing relation to the upper halved portion of the fruit A and the tines 14 and 16 are respectively embedded or anchored in the skin or rind B of the fruit so that the hood shaped body 10 projects upwardly above the fruit to catch and prevent the spattering of the juice when eating the same. In view of the fact that the hood shaped body 10 and the strips 12 are flexible to a degree, it is obvious that the device may accommodate fruits of various sizes. It is to be further understood that within the scope of the invention the shield may be made of any desired material, but it is preferable to make the same of paper or other cheap material, whereby the same may be thrown away after a single use. The shield body and the strips 12 and 15 may be used to bear advertising matter. [See Figs. 1 and 2.]

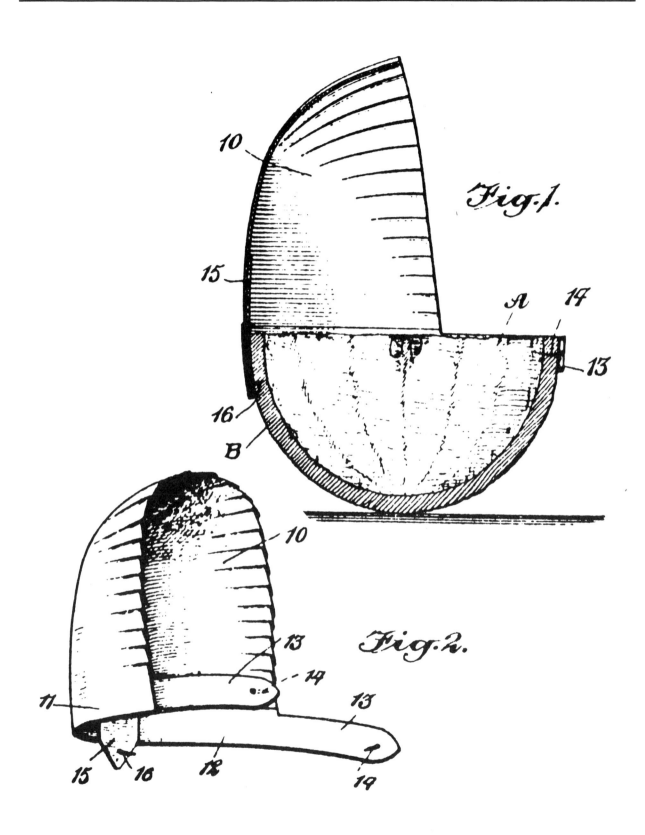

Bedwetting Alarm
Patent No. 1,772,232 (1930)
Jesse S. Van Guilder of Red Wing, Minnesota

The primary object of the invention is to provide means for sounding an alarm when urination starts so as to awaken the sleeper before the bladder is emptied and thereby enable him or her to avoid material wetting of the bed.

In operation, the filling material [see Figs. 1 and 2] is normally dry and insulates the sheets 7 and 8 from each other so as to prevent sounding of the bell 17 even though the switch 15 is closed. When the sleeper begins to urinate, the urine runs through the casing 5 and conductor sheet 8 and saturates a portion of the filling material 6 entirely through to the sheet 7. The wetted portion of the material 6 is thus rendered conductive to electrically connect the sheets 7 and 8 to operate and awaken the sleeper before his or her bladder is emptied. By turning the switch 15 off, the alarm may be rendered silent, and by constructing the casing so that it may be opened, the wet filling material may be replaced by dry material to condition the device for re-use.

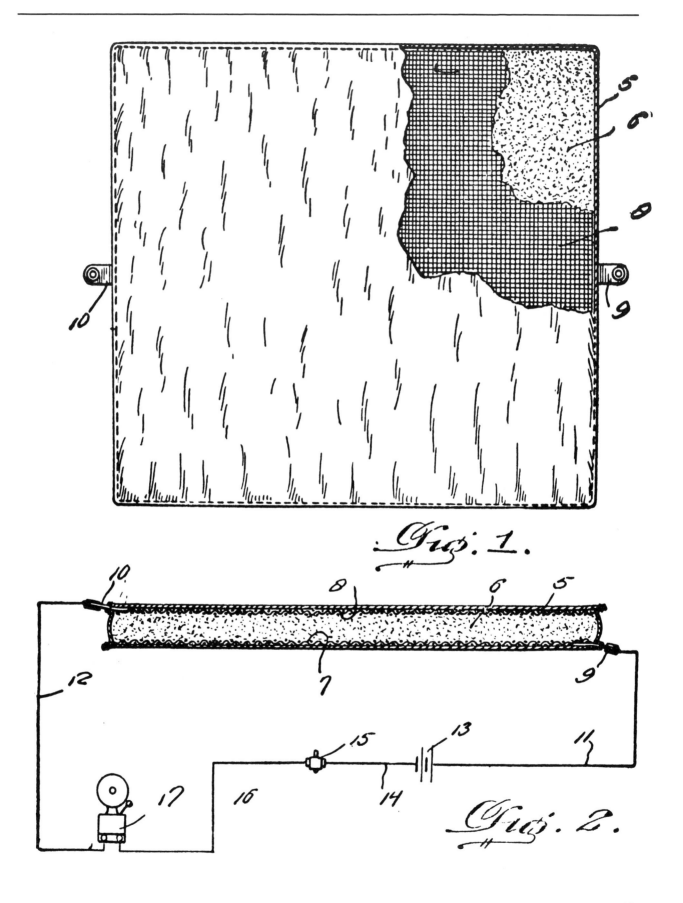

Wind Bag
Patent No. 2,149,053 (1939)
Thomas Lloyd Hollister of Miami, Florida

This invention relates to a device for receiving and storing gas formed by the digestion of foods.

An object of the invention is the provision of a device for collecting and storing gas formed in the digestive tract, said device being removably suspended from the body and provided with a nipple having shielded perforations to permit gases to enter a storage chamber.

A further object of the invention is the provision of a device for taking care of the excess gas under pressure in the alimentary tract whereby strains are relieved.

A still further object of the invention is the provision of a device having a perforated nipple adapted to be received by one end of the alimentary tract through which gas is adapted to pass for storage in a chamber remote from the nipple, the perforations in the nipple being shielded to prevent clogging thereof with a muffler between the nipple and the storage for baffling the gases when under pressure.

Referring more particularly to the drawings [see Figs. 1 and 2], 10 designates a strap or belt adapted to embrace the waist of the body and from which depend straps 11 and 12. The lower end of the strap 11 is provided with a loop 13 adapted to receive an eye 14 secured in any approved manner to an intermediate portion of a conduit or tube 18. The strap 12 is provided with a loop 16 adapted to receive an eye 17 projecting from a collar or flange 18 secured to or formed integrally with the upper portion 19 of the tube 45.

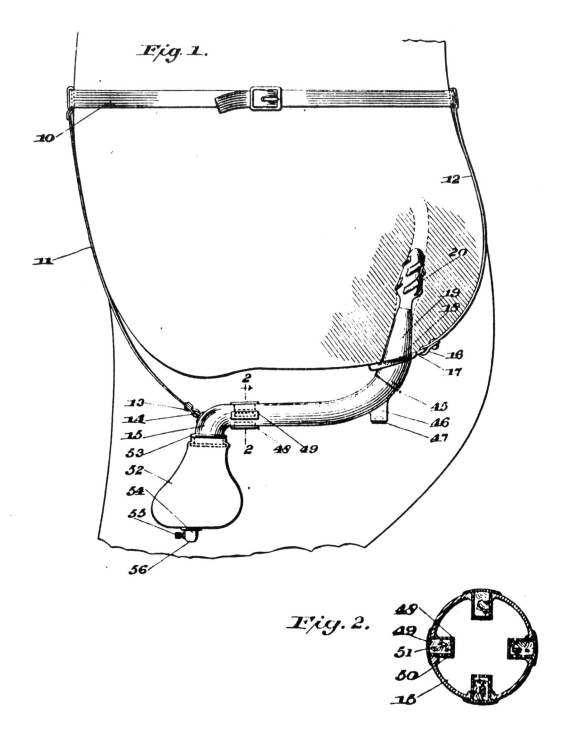

Bird Diaper
Patent No. 2,882,858 (1959)
Bertha Dlugi of Milwaukee, Wisconsin

The practice of maintaining parakeets and similar birds as household pets has gained in popularity. The popularity of this type of bird is partially due to its unique characteristics which permit it to be granted the freedom of the house, and they are usually allowed to fly about all or some of the rooms of the home at times. A distinct disadvantage in this practice is that these birds cannot normally be house trained as other pets are, and their excremental discharge is frequently deposited on household furnishings when they are at liberty, creating an unsanitary condition.

According to the present invention the improved sanitary garment comprises a triangular patch of material adapted to be supported about the crissum of the bird for receiving its excremental discharge. The patch is preferably in the form of an isosceles triangle that is supported by a harness formed by strips of tape [Figs. 1–3]. One of these strips of tape extends over the base of the bird's tail and another strip of tape is attached to the apex of the triangle, extending upwardly therefrom along the breast of the bird for attachment to a neck band that encircles the neck of the bird. The harness is completed by another strip of tape extending along the bird's back, being attached at one end to the neck band and at its opposite end to the tape disposed on the top of the tail of the bird. Thus, the harness firmly supports the patch of material in position to perform its function. The several strips of tape are of a relatively narrow width, and after the harness has been placed upon the bird the bird will invariably ruffle its feathers causing them to overlie the strips of tape and thereby effectively conceal them. A length of elastic is provided along the base of the triangular patch to draw this portion of the patch together for the purpose of yieldably drawing the slack about the base of the bird's tail and at the same time causing the central portion of the body of the patch to sag and thereby create a pouch for receiving the excremental discharge.

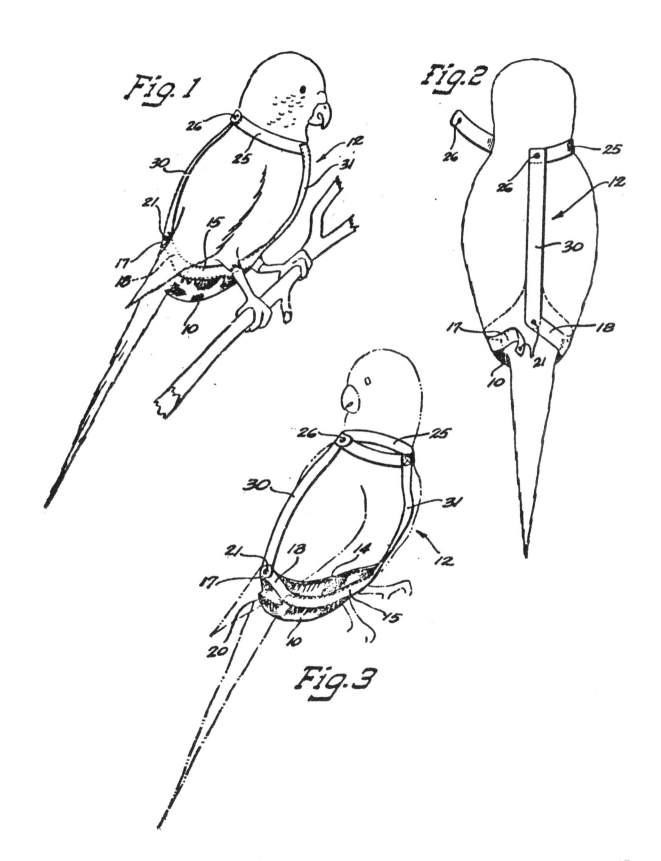

Eyeglass Wiper
Patent No. 2,888,703 (1959)
Klara Karwowska of Calgary, Alberta, Canada

The object of the invention is to provide an eyeglass wiper mechanism which will effectively keep eyeglasses clean of rain, steam, or the like.

Another object of the invention is to provide a pair of eyeglasses which are provided with a means for keeping the lenses free or clean so that even when it is raining, snowing or the like, the person wearing the glasses will be able to readily actuate the wipers so as to keep the lenses clean for clear visibility.

Referring in detail to the drawings [Fig. 1–4], the numeral 10 designates the eyeglass frame which can be made of any suitable material, and a pair of temples or side pieces 11 are hingedly connected to the ends of the frame 10 through the medium of hinge pins 12. Mounted in the frame 10 is a pair of lenses 13.

The present invention is directed to a wiper means for maintaining the lenses 13 clean or clear of steam, rain, snow, or other foreign matter, and the wiper mechanism of the present invention includes a source of electrical energy such as the battery 14 which may be secured to the frame 10 in any suitable manner.

Also carried by the frame 10 is an electric motor 15 which may be provided with a manually operable switch or control button 16 whereby upon actuation of the switch 16, the motor 15 can be energized. The electric wire 17 may lead from the motor 15 to the battery 14.

There is further provided a pair of moveable wiper blades 18 and 20.

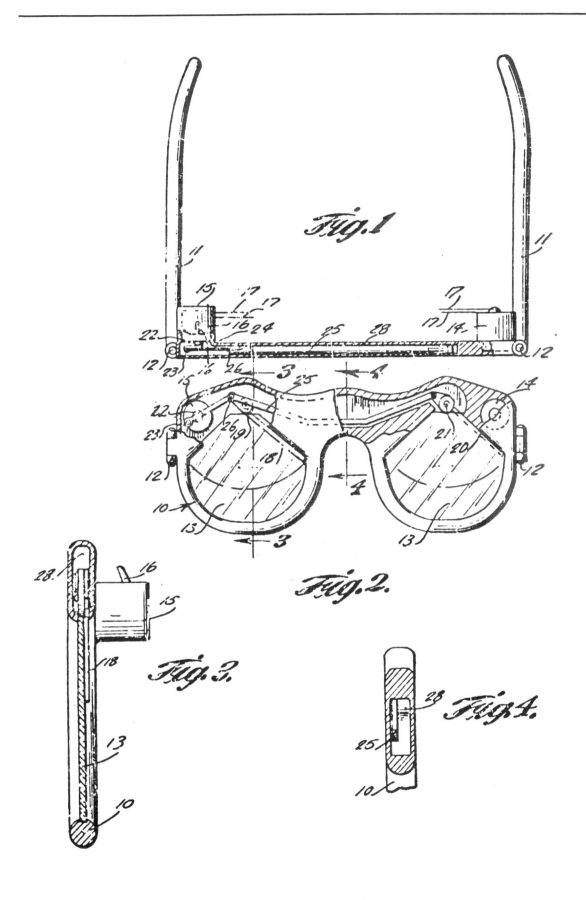

Religious Soap
Patent No. 3,936,384 (1976)
Yancey Williams of New York, New York

The objects and features of the invention may be understood with reference to the following description of an illustrative embodiment of the invention:

Figs. 1–3 illustrate the bar of soap which is embossed on the top face with the markings 11 of a crucifixion scene, and on the opposing face with the markings 13 of a prayer. A side face may be embossed with markings 12 of the designer or producer of the soap bar 10.

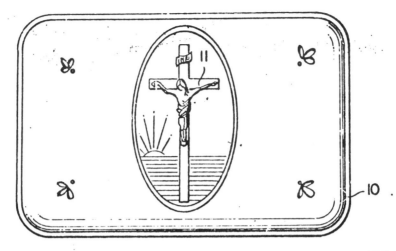

FIG.1

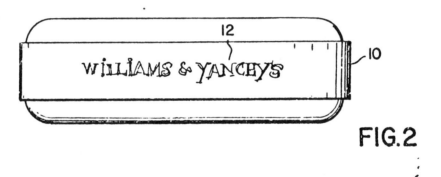

FIG.2

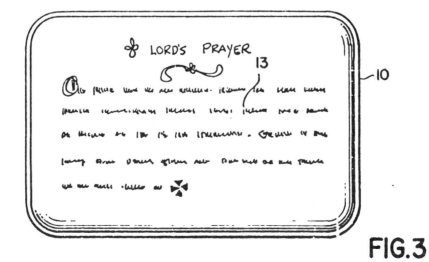

FIG.3

THREE
Rube Goldberg Is Alive and Well

Professor Lucifer Gorgonzola Butts (alias Rube Goldberg) had an inordinate love of levers, gears, pulleys, and knobs. So do many of the inventors on the pages that follow.

There are some obvious and phenomenal success stories among them, entrepreneurs who were able to combine their mastery of mechanics with a logical purpose and thereby achieve a marketable patent. Many were concerned with conveying the printed word. Four hundred and fifty years after Johann Gutenberg invented the printing process in Germany in 1440, Ottmar Mergenthaler of Baltimore was awarded a patent for his amazingly complex typesetting machine. The drawing on page 58 is one of twenty-three detailed illustrations he submitted. There were also the typewriter in 1868 by Christopher Sholes of Milwaukee, which created many new jobs for working women (p. 56); the fountain pen by Lewis Waterman of New York, in 1884; and over a half-century later the first ballpoint, by Milton Reynolds of Chicago, in 1938.

One of the earliest American patents was awarded to Eli Whitney in 1794 for his simple box-and-comb device for extracting cottonseeds from fiber (see p. 118). The picture camera was another ingenious contraption in a box: Louis Daguerre's prototype (1839) brought together the sciences of optics, physics, and chemistry. So did George

> If it is not true, it is very well invented.
> *Giordano Bruno,*
> Degli Eroici Furori

Eastman's Kodak in 1880 (which carried with it the slogan "You press the button, we do the rest"), Chester Carlson's copying machine in 1942 (p. 70), and Edwin Land's Polaroid in 1948 (p. 72).

Gears and pulleys likewise played a big part in the development of large cities and their skyscrapers. Until 1840 few buildings were higher than three or four stories. The crane was patented in 1839 by William Otis of Philadelphia (p. 52). Then came the safety elevator, which Charles Otis unveiled to the world at the Crystal Palace Exposition in New York in 1853 (p. 54). Charles Seeberger's escalator appeared in 1900.

Professor Butts would have been proud of John Larson, who invented the polygraph machine in 1921 while a medical student at the University of California. Until then the most popular ways of detecting lies relied upon the primitive "dry mouth" tests: The ancient Chinese forced a suspect to chew and spit out a handful of rice (if it was still dry, the suspect was guilty), and the early English defendant had to swallow a "trial slice" of bread and cheese to prove his innocence.

The inventors of the various alarm beds and combination rockers depicted in this chapter, on the other hand, might have accused Rube Goldberg of plagiarism.

Crane Excavator
Patent No. 1,089 (1839)
William S. Otis of Philadelphia, Pennsylvania

The drawing which accompanies this specification represents the crane, which I employ in excavating and with which I am at present operating, but I do not mean to confine myself to the exact construction and arrangement of this part, as it may be varied in its details, while the same effects may be produced by it. This crane is shown in Figure 1 as stationed upon a movable carriage, or car, and is situated on a temporary railway, and is intended to be so constructed that the load of earth taken off by the scraper may be raised by it, while the turning of the crane also to the point where the load is to be dumped, whether into cars or boxes on the road, or elsewhere, is controlled and regulated with great precision.

In the accompanying drawing, *A* represents the scraper suspended from the crane, by means of the ordinary tackle; this tackle serving to raise it in the act of excavating, and to lower it when required. The end *B* of this scraper is hinged to it, and has a bolt which keeps it in place when desired, or allows it to be opened and discharge its contents from that end.

In using this machine it is of course to be moved forward as the excavation proceeds, and this may be done either upon a temporary railway, or upon the ground, as may be found most convenient. When brought into use it is necessary that it should be braced laterally, and this I effect by means of a projecting bench *P*, on each side of the carriage, sustained by a brace *Q*, and having at their ends a screw *R*, which screw is made to bear upon blocking placed under it for that purpose. By this provision all lateral motion is effectually checked.

Thus fully described [are] all those parts of my excavating crane which I believe to be new and the manner in which these improvements are connected with the ordinary gearing, by which cranes are or may be operated upon by any motive power, such as steam, animals, or men.

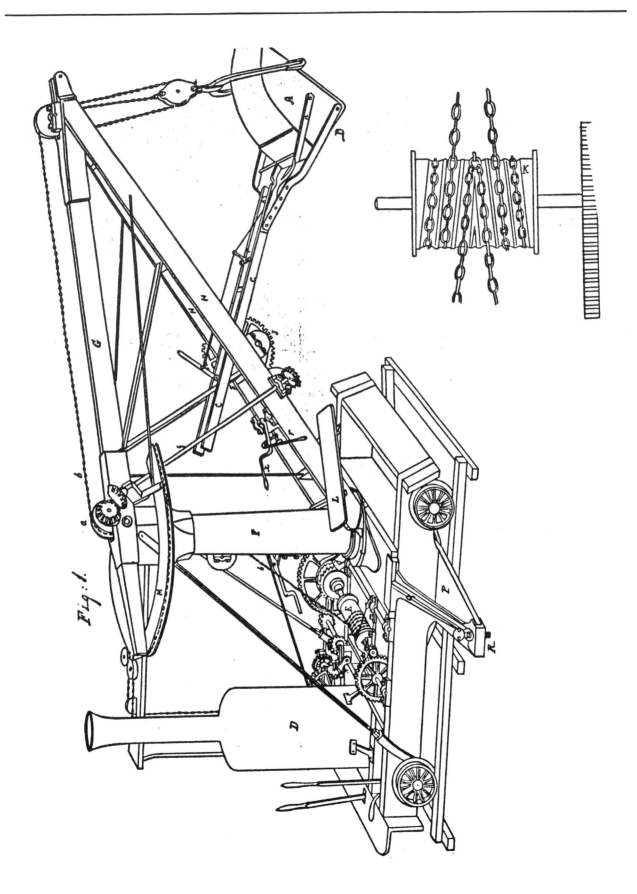

Elevator
Patent No. 113,555 (1871)
Charles R. Otis and Norton P. Otis of Yonkers, New York

Improvements to the passenger elevator were concerned primarily with safety. The one here consisted of a novel means of producing the operation of the stop-pawls in the ratchets provided on the upright guides between which the car, cab, or platform works; and of producing the operation of a friction-brake to provide against accident in cases of the breaking of the lifting-rope and in cases where the said rope does not break, as, for instance, where its weight in slackening might be sufficient to prevent the action of the stops; and to provide generally against accident consequent on any tendency to the too rapid descent of the cab, car, or hoisting-platform, or in case of the slackening of the lifting-rope from whatever cause; such means to be used either alone or in addition to other means which may be employed to effect the said stoppage, or check the descent of the cab, car, or platform under similar circumstances.

The invention likewise includes a combination of a weight and a brake-lever with the stop-lever of the engine, for the purpose of arresting the driving-power of the apparatus in case of the hoisting. [See Figs. 1 and 2.]

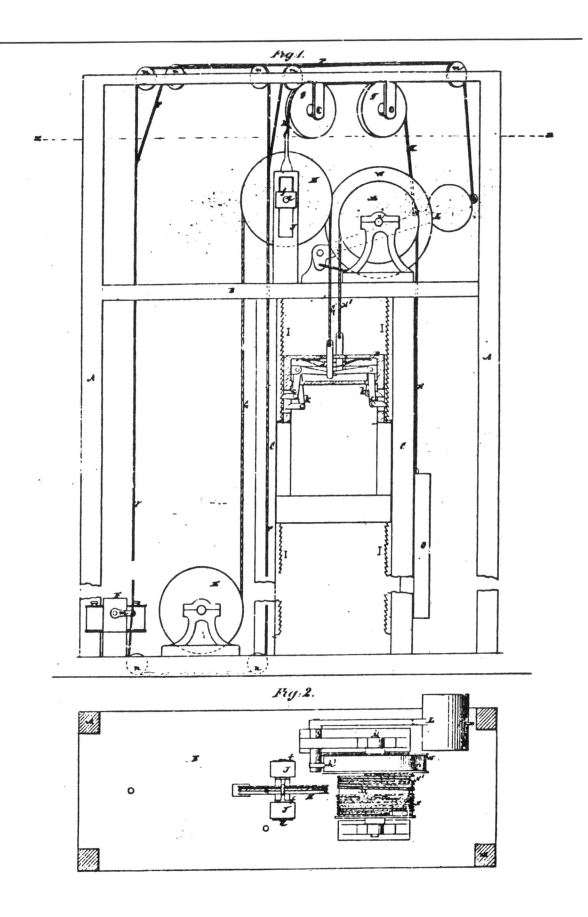

Typewriter
Patent No. 79,265 (1868)
C. Latham Sholes, Carlos Glidden, and Samuel W. Soule, all of Milwaukee, Wisconsin

This invention is of improvements to an invention of a type-writing machine, an application for a patent for which we filed October 11, 1867. Its features [see Figs. 1, 2, 3 and 9] are a better way of working the type-bars, of holding the paper on the carriage, of moving and regulating the movement of the carriage, of holding, applying, and moving the inking-ribbon, a self-adjusting platen, and a rest or cushion for the type-bars to follow.

Make a case *A*, about two feet square, four to six inches deep, or of any requisite dimensions, of material and finish to one's taste. In the cover cut a circle. Make a circular annular disk *B* of any hard tough material (we use and prefer brass) four to five inches in diameter, or any required size, with a circle or hole in the center, one to one and a half or more inches in diameter, with the outer edge or periphery one-half to three-fourths of an inch or more thick, and the inner edge or circumference of the central circle two-eighths to three-eighths of an inch or more thick, with the top side planed level and smooth and the bottom side beveled, if preferred, from the outer to the inner edge with as many radial slots or grooves as types to be used cut in the bottom side from the central circle to the periphery, and deep to within an eighth of an inch of the top, less or more, with slots in the outer edge or periphery one-half to three-fourths of an inch or more deep toward the central circle to meet and fit exactly the radial grooves, and with a groove for pivot-wire cut in and circumscribing the periphery, as shown in Fig. 1.

Of any suitable material (we use and prefer steel) make as many type bars or hammers *o* as types to be used or slots in the disk. Pivot the outer ends of the type-bars in the slots in the outer edge by a wire laid in the groove in the periphery circumscribing the disk. On the upper sides of the inner ends of the type-bars cut in relief the types to be used. Make all the type-bars of the exact length of the radius of the circle of the disk, so each type on the inner ends, when thrown up into the radial grooves, will strike against the central point.

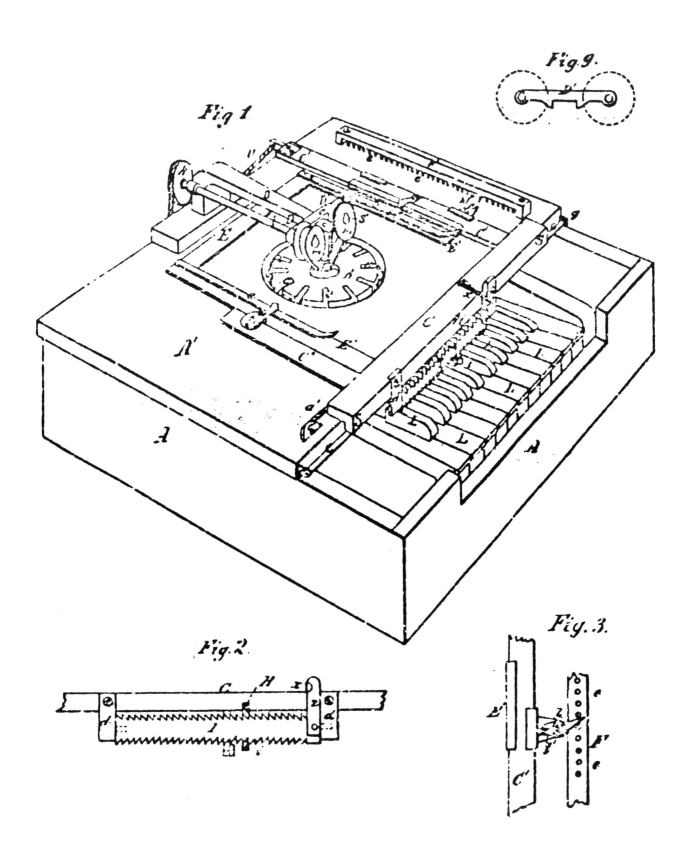

Linotype Machine
Patent No. 436,532 (1890)
Ottmar Mergenthaler of Baltimore, Maryland

The present invention relates to a machine for producing what are known in the art as "linotypes"—that is to say, bars for types, each having on one edge the characters necessary to print an entire line of a page or column.

The present machine contains as its fundamental feature a series of independent female type or matrices and a series of space-bars [see Fig. 1]. The matrices and space-bars properly assorted are contained in magazines or holders. A series of finger-keys representing the respective characters and the space-bars act, in connection with suitable composing mechanism, to assemble the matrices in line in the order in which they are to appear in print, and also to introduce the spaces at suitable points in the line. After the matrices and spaces for an entire line are assembled a shifting mechanism transfers them to the front of a mold, the internal form and dimensions of which correspond with those of the required linotype. While the line is thus located in front of the mold, the internal form and dimensions of which correspond with those of the required linotype. While the line is thus located in front of the mold suitable clamping devices act thereon and the space-bars are advanced through the line, so as to "justify" the same. A melting-pot containing constantly a large body of molten metal is arranged to close the mold on the rearside, and at the proper moment a pump in this pot acts to deliver the molten metal into the mold, where its solidifies and produces a linotype, which receives on its edge the impression of the matrices at the front. As soon as this casting operation is completed the clamps lift the matrices from the front and the space-bars are then separated mechanically from the matrices and return to the magazine from which they started, while the matrices are carried to a distributing mechanism at the top of the machine, where they are assorted and returned to the upper ends of the appropriate magazine-tubes. After the removal of the matrices therefrom the mold makes a partial revolution and an ejecting device delivers the linotype therefrom to a galley or receiver....

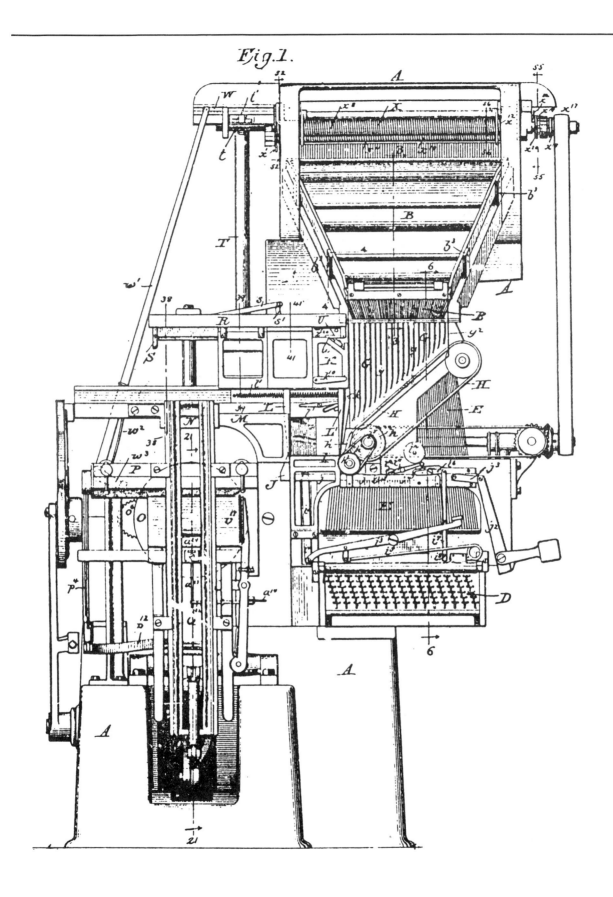

Coffin with Escape Hatch
Patent No. 81,437 (1868)
Franz Vester of Newark, New Jersey

The nature of this invention consists in placing on the lid of the coffin, and directly over the face of the body laid therein, a square tube, which extends from the coffin up through and over the surface of the grave, said tube containing a ladder and a cord, one end of said cord being placed in the hand of the person laid in the coffin, and the other end of said cord being attached to a bell on the top of the square tube [see Figs. 1–4], so that, should a person be interred ere life is extinct, he can, on recovery to consciousness, ascend from the grave and the coffin by the ladder; or, if not able to ascend by said ladder, ring the bell, thereby giving an alarm, and thus save himself from premature burial and death.

The operation of my invention is as follows: The supposed corpse being laid in the body A of the coffin, and the cord K placed in the hand of the corpse, the cord is next drawn through the tube C and attached to the bell I, and the tube C placed in the base D, on the lid of the coffin. The coffin is now lowered into the grave, and the grave filled up to the air-inlets FF. Now, should the person laid in the coffin, on returning to life, desire to ascend from the coffin and the grave to the surface, he can do so by means of the ladder H; but, if too weak to ascend by the ladder, he can pull the cord in his hand, and ring the bell I, giving the desired alarm for help, and thus save himself from premature death by being buried alive. Should life be extinct, the tube C is removed, the door L closed, and the tube used for a similar purpose.

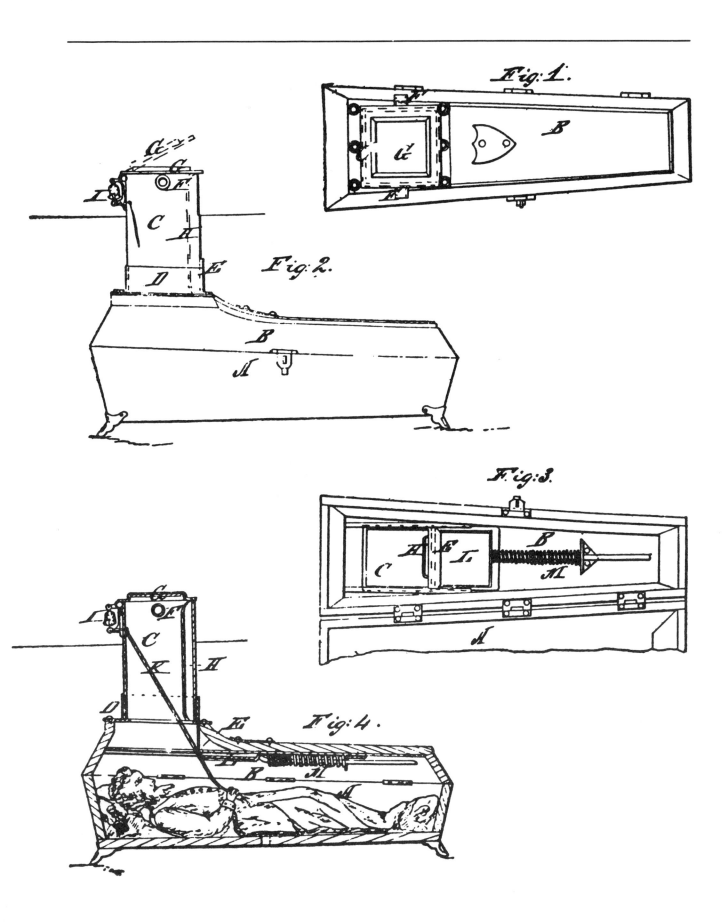

MOUSETRAPS AND MUFFLING CUPS

Alarm Bed (I)
Patent No. 256,265 (1882)
Samuel S. Applegate of Camden, New Jersey

Ordinary bell or rattle alarms are not at all times effective for their intended purpose, as a person in time becomes so accustomed to the noise that sleep is not disturbed when the alarm is sounded.

The main aim of my invention is to provide a device which will not be liable to this objection.

I suspend a light frame in such a position that it will hang directly over the head of the sleeper, the suspending-cord being combined with automatic releasing devices, whereby the frame is at the proper time permitted to fall into the sleeper's face.

[In Figs. 1–5] A represents the frame, which consists of a central bar, a, having on each side a number of projecting arms, b, the whole being made as light as is consistent with proper strength. From each of the arms b hang a number of cords, d, and to the lower end of each of these cords is secured a small block, e, of light wood, preferably cork . . . the only necessity in being that when it falls it will strike a light blow, sufficient to awaken the sleeper, but not heavy enough to cause pain.

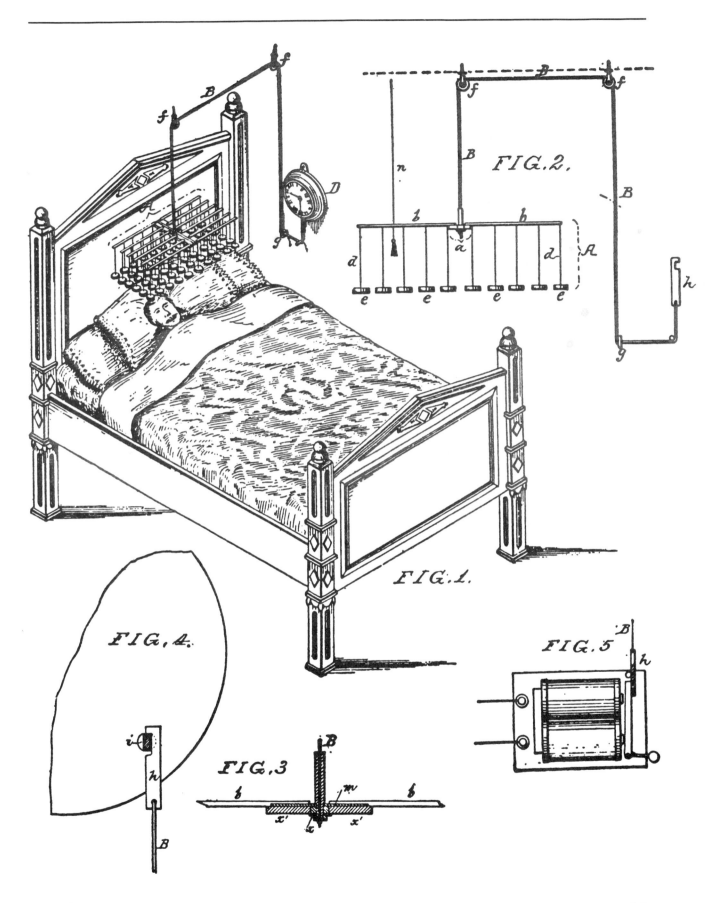

Alarm Bed (II)
Patent No. 643,789 (1900)
Ludwig Ederer of Omaha, Nebraska

The object of my invention is to provide a bed to be used more particularly in connection with hothouses and conservatories and is arranged in such a manner that should the steam in the heating-pipes fall below a certain pressure it will tilt the bed to awaken the sleeper having charge of the heating plant.

In the accompanying drawings I have shown in Fig. 1 an alarm-bed embodying my invention, showing the bed as tilted in dotted lines, while Fig. 2 shows an enlarged detail, detached, of a diaphragm by means of which the bed is operated.

Now, as the steam-pressure in the boiler and the connected pipes F decreases, the diaphragm D gradually sinks downward, following the diminishing pressure with the pipe system until the roller p rides off of the rail and below the bar e, when the weight of the sleeper upon the bed insures the tilting of the bed, so that the sleeper will slide or roll off, thus reminding him that the steam within the pipe system is below a certain point, endangering the life of the plants within the greenhouse. After the operator has replenished his fire the steam of course promptly carries the diaphragm up again, so that the bed, which had been in a horizontal position, by means of the spring S will be locked and ready to receive the occupant, and it will be held in a locked position until the steam in the boiler again falls below a certain point.

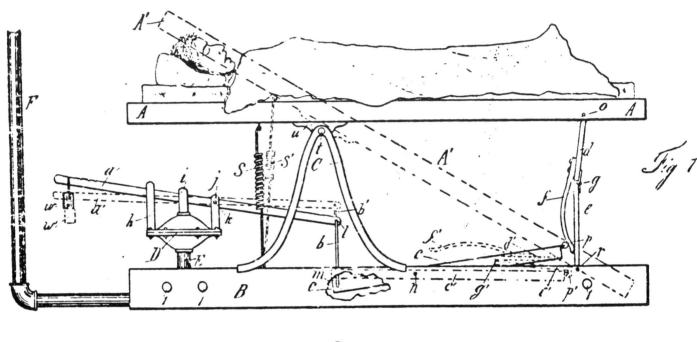
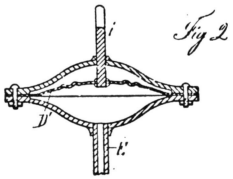

Rocking Chair Fan
Patent No. 92,379 (1869)
Charles Singer of South Bend, Indiana

This invention relates to improvements in the construction of rocking-chairs, with air-blowing attachments, having for its object to provide a stand or base for the support of a bellows, with tracks or rails, on which the rockers, which are fixed close to the seat, may work, instead of on the floor; also, to provide an arrangement whereby the parts may be readily detached for storage or packing in compact form; and also an improved arrangement of parts, whereby the bellows is operated, all as hereinafter specified.

This stand, with elevated rails, protects the rockers against rocking on small children crawling on the floor, or strings scattered thereon. It also provides for rocking the chair with the same ease, when sitting on the ground; and it also serves as a support for a bellows D [see Figs. 1 and 2], whereby the occupant may, by the act of rocking, impel a current of air upon himself, through a flexible tube, E, which may be directed to any part, as required.

The top of this bellows is connected by a bent bar, F, to the stand A, so as to be held in a fixed position, while the lower part is connected by a similar bent bar, G, to the bottom of the chair, so as to be moved up and down with it, to impel the air.

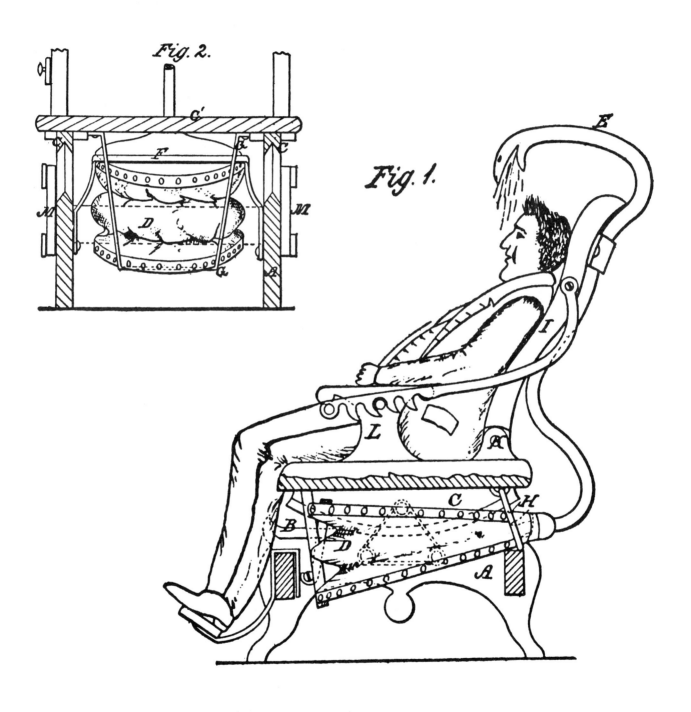

Rocking Chair Butter Churn
Patent No. 1,051,684 (1913)
Alfred Clark of Corinth, Maine

This invention relates to a rocking chair which churns butter as it rocks.

In the accompanying drawing, Figure 1 is a side elevation of the churn operating mechanism, and Figure 2 is a sectional view. The device consists of a shaft mounted for rotation, a ratchet wheel fast on said shaft, rocking arms pivotally mounted on said shaft, arranged respectively on opposite sides of the ratchet wheel and provided with pawls to engage opposite sides of the ratchet wheel, in combination with a chair rocker, a support therefor, and rods connecting the chair rocker to the respective rocking arms.

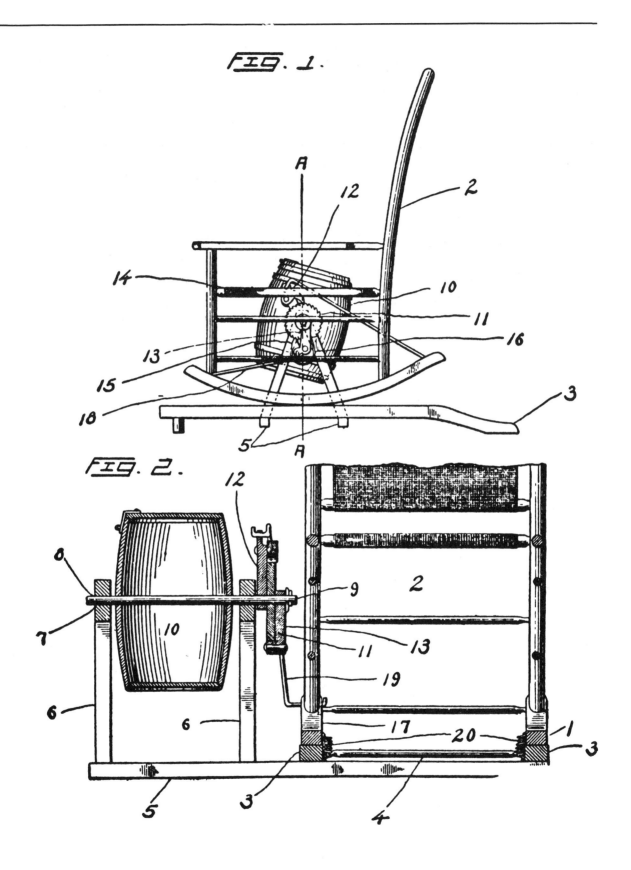

Copying Machine
Patent No. 2,297,691 (1942)
Chester F. Carlson of Jackson Heights, New York
(assignor to Haloid Company, later Xerox)

Figure 1 is a section through a photographic plate according to my invention and illustrates a preferred method of applying an electric charge to it preparatory to photographic exposure;

Figures 2, 2a and 2b illustrate three methods of photographically exposing the plate;

Figures 3 and 4 show a method of developing the electrostatic latent image produced on the plate by the preceding steps;

Figure 5 shows a method of transferring the image to a sheet of suitable material such as paper;

Figures 6 and 7 illustrate methods of fixing the image onto the sheet;

Figure 8 illustrates a modified means for charging and exposing the photographic plate;

Figure 9 shows another method of developing the image; and

Figure 10 is an enlargement of a half-tone produced by the process.

Certain advantages of my process over ordinary photographic methods will be apparent. In the first place the process yields a direct positive copy, instead of a negative. That is, upon exposure to the original and then dusting with a black or colored powder and transferring to a white sheet of paper, the areas which appeared dark on the original will be reproduced as black or colored areas on the copy, and the areas which were white on the original will also be white on the copy.

Another advantage is that the process yields directly readable copies of written or printed matter with the use of an ordinary camera lens or by contact printing with the printed side against the sensitive plate, rather than yielding a mirror image. This makes the process especially well adapted to direct reproduction of printed matter, drawings, typewritten matter and the like. For instance, if it is desired to reproduce a typewritten letter it may be copied either with the camera, as shown in Figure 2, or by contact printing as shown in Figure 2a, and the finished copy 36 (Figure 6 or 7) will be an exact black-letters-on-white-background duplicate of the original letter.

An outstanding advantage of the process described herein is its simplicity and rapidity. It is a matter of only a few seconds, to make a complete permanent copy of any original. No complex chemical development process is required.

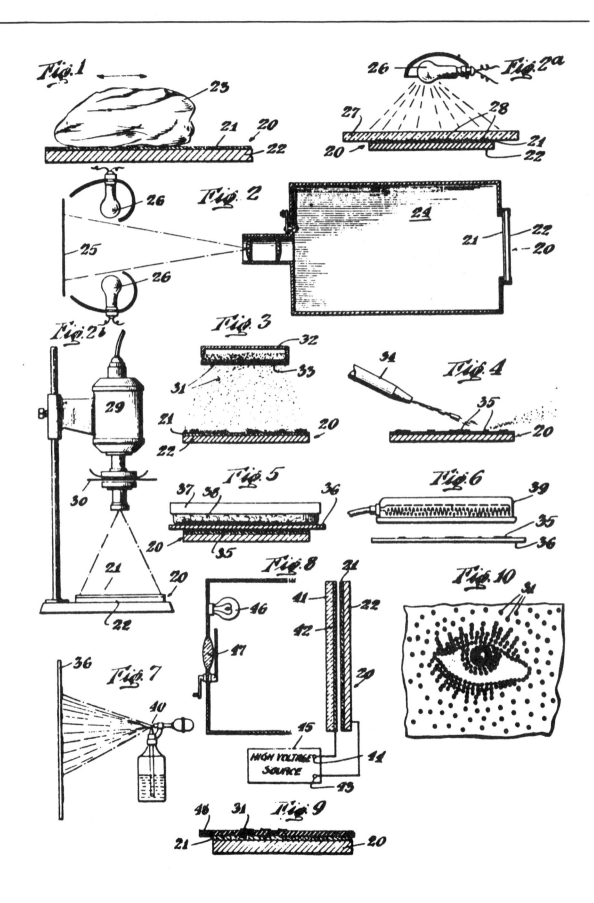

Polaroid Camera
Patent No. 2,435,720 (1948)
Edwin H. Land of Cambridge, Massachusetts

This invention relates to photography, and more particularly to apparatus for bringing photographically exposed photosensitive film material and individual sheets of another preferably nonphotosensitive material into operative assembly and for spreading a processing liquid within the assembly of materials by providing a predetermined compression thereof.

An object of the present invention is to provide novel photographic apparatus for positioning, in operative assembly, a photographically exposed film material, an individual sheet of another material, and a fluid releasably confined in a container attached to one of the materials, and for applying a predetermined pressure to the assembly for at least spreading the fluid between the aforesaid materials to process said photographically exposed film material and preferably to provide said assembly with a positive photographic image.

Figure 1 is a somewhat schematic side-elevation view, partly in cross section and with parts broken away, of one form of the novel camera apparatus of the invention....

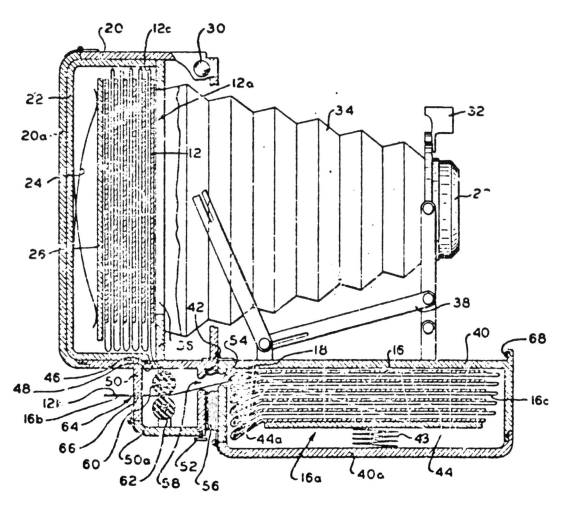

FIG. 1

FOUR
Necessity, the Mother

"You cannot escape necessities," said Seneca, "but you can conquer them." And indeed we conquer with vengeance. Inventors have understandably applied themselves first (and most profitably) to the improvement of food, clothing, and shelter, and few would argue that we are eating, dressing, and living more comfortably and conveniently now than in previous generations.

We eat better primarily because of packaging and refrigeration. Newer methods of food preservation were developed in France for Napoleon's armies. (Nicolas Appert, a Parisian chef, won a 12,000-franc prize in 1809 for discovering how to keep food fresh longer by using glass jars and corks.) The first sealed canisters were introduced by a London merchant, Peter Durand, in 1830. It took thirty years longer, however, for someone to invent what we know as the can opener—prior to 1860 consumers were instructed to use hammers and chisels. Jacob Perkins, an American living in England, won a British patent in 1834 for a means of cooling liquids to produce ice; by 1850 refrigerators were being patented in the United States. People began to start their days differently with the coming of cream separators (1879) and breakfast cereals (1897). The latter were introduced by Harvey and Will Kellogg, who dispensed cornflakes at their health spa in Battle Creek, Michigan. One of their early guests, Charles William Post, was inspired

What we do we must, and call it by the best names.
Ralph Waldo Emerson, Conduct of Life

to create Grape Nuts Flakes. Since then all manner of cooking devices have received patents, the most recent revolutions being microwave ovens (1948) and nonstick pans (1955).

The making of cloth was refined most dramatically during the eighteenth century, with numerous inventions from the spinning wheel to the cotton gin (see p. 118). Elias Howe, of Spencer, Massachusetts, produced fourteen sewing machines in 1850; the next year Isaac Singer of New York patented an improved version—and both men became rich. Elastic and rubber fabrics were introduced in the 1820s, and man-made fibers were exhibited as early as 1883 (though nylon didn't appear until 1936).

The last two centuries have also seen the greatest advances in building materials and techniques, with the development of everything from prestressed concrete (1854) to plywood (1874) to plastics (1889), and from the modern crane (1839) to the safety elevator (1953) to the first skyscraper (the forty-one story Singer Building, erected in 1907).

What it is that separates folly from necessity often remains a grand mystery, though the common sense of hindsight enables us to explain why the world has given the cold shoulder to the foot warmer (p. 82) while it has enthusiastically latched on to the zipper (p. 90) and its progeny.

Sewing Machine
Patent No. 4,750 (1846)
Elias Howe, Jr., of Cambridge, Massachusetts

In sewing a seam with my machine two threads are employed, one of which thread is carried through the cloth by means of a curved needle, the pointed end of which is to pass through said cloth. The needle used has the eye that is to receive the thread within a small distance—say, an eighth of an inch—of its inner or pointed end. The other or outer end of the needle is held by an arm that vibrates on a pivot or joint pin, and the curvature of the needle is such as to correspond with the length of the arm as its radius. When the thread is carried through the cloth, which may be done to the distance of about three-fourths of an inch, the thread will be stretched above the curved needle, something in the manner of a bow-string, leaving a small open space between the two. A small shuttle carrying a bobbin filled with silk or thread is then made to pass entirely through this open space between the needle and the thread which it carries, and when the shuttle is returned, which is done by means of a picker-staff or shuttle-driver, the thread which was carried in by the needle is surrounded by that received from the shuttle, and as the needle is drawn out it forces that which was received from the shuttle into the body of the cloth, and as this operation is repeated a seam is formed which has on each side of the cloth the same appearance as that given by stitching, with this peculiarity, that the thread shown on one side of the cloth is exclusively that which was given out by the needle, and the thread seen on the other side is exclusively that which was given out by shuttle. It will therefore be seen that a stitch is made at every back-and-forth movement of the shuttle. The two thicknesses of cloth that are to be sewed are held upon pointed wires which project out from a metallic plate, like the teeth of a comb, but at a considerable distance from each other—say three-fourths of an inch, more or less—these pointed wires sustaining the cloth and answering the purpose of ordinary basting. The metallic plate from which these wires project has numerous holes through it, which answer the purpose of rack-teeth in enabling the plate to be moved forward by means of a pinion as the stitches are taken. [See Figs. 2 and 5.]

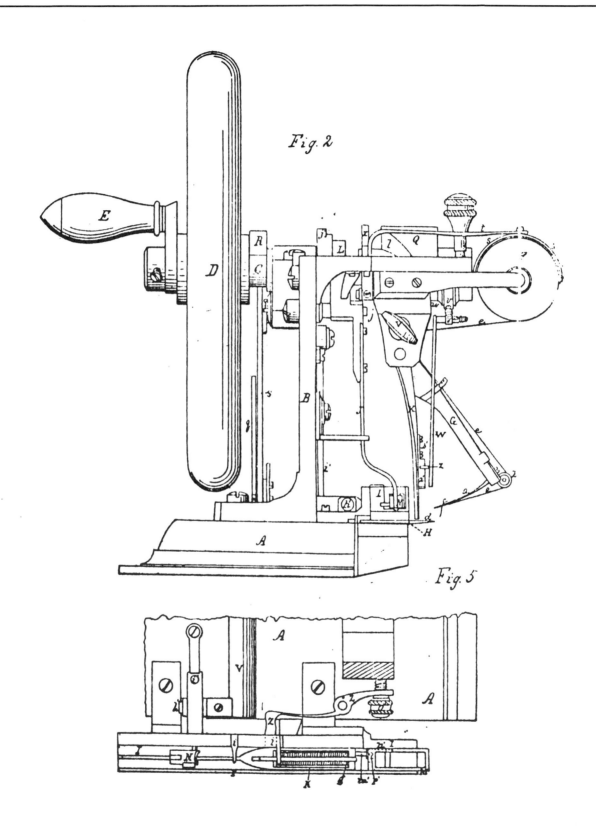

Umbrella Skirt
Patent No. 44,482 (1864)
Elizur E. Clarke of New Haven, Connecticut

My invention relates to an improvement in the common umbrella whereby that useful article is rendered a more perfect protector from storms; and it consists in attaching to a common umbrella under and at a little distance in from the edge a skirt of any kind of fabric, which extends entirely around the umbrella and down nearly to the feet of the person carrying it [Figs. 1 and 2].

I attach to the ribs of a common umbrella, *A*, hooks, loops, buttons, or their equivalent, to which to secure the skirt. I construct the skirt *B* from a piece of fabric (preferring water-proof cloth) of proper width to extend around the umbrella, and in length so as to extend nearly to the feet of an ordinary-sized person—say about four feet. I hang the said skirt by its upper edge to the ribs of the umbrella by means of the aforesaid hooks, loops, or buttons. The two edges which then hang perpendicular from the umbrella I join by buttons, as shown, or may be stitched up part of the distance or otherwise closed, so as to leave an opening at the top for the person walking inside to look from, as seen in Fig. 1.

Suspenders
Patent No. 121,992 (1871)
Samuel Clemens (Mark Twain) of Hartford, Connecticut

The nature of my invention consists in an adjustable and detachable elastic strap for vests, pantaloons, or other garments requiring straps, as will be hereinafter more fully set forth.

Figure 1 represents an elastic strap made in two parts, C and C', each part being provided at one end with one or more button-holes, c, and the other ends of said parts connected by a buckle, c'. Fig. 2 represents an elastic strap made in two parts, D and D', each part being provided at one end with one or more button-holes, d. The part D has at the other end several rows of holes, d^1 d^1, into which hooks d^2 at the other end of the Part D' are to be inserted. This end of the part D' may be lengthened or shortened by means of a slide, d^3, as shown.

The vest, pantaloons, or other garment upon which my strap is to be used should be provided with buttons or other fastenings on which the strap is to be detachable and adjusted. When changing garments the strap may readily be detached from one and put on another.

The advantages of such an adjustable and detachable elastic strap are so obvious that they need no explanation.

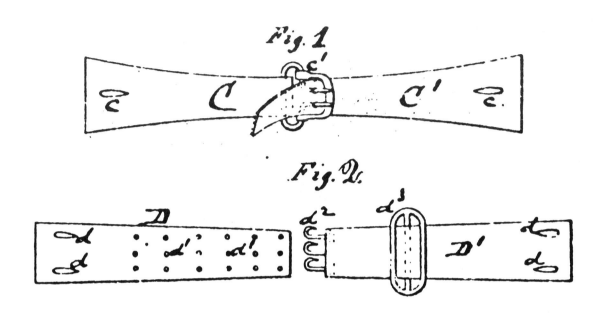

Foot Warmer
Patent No. 186,962 (1877)
William Tell Steiger of Howard County, Maryland

It is a well-established fact that our lungs constitute the laboratory of nature, within which—by a condensing process—animal heat is generated, and afterward conveyed and distributed to other portions of our bodies by the action of the heart and circulation of the blood; that, for mechanical reasons, the supply to the extremities—the hands and feet—on account of their distance from the center of heat, is more or less deficient, and, consequently, they suffer most when exposed to severe outward cold; the feet, especially, by reason of their immediate contact, in winter weather, with cold floors, as in railroad-cars and other vehicles, and with the frozen ground and icy sidewalks.

Now, I find, by personal experiment, that by breathing for a short time on the bulb of a thermometer I am enabled to raise the mercury to 88° Fahrenheit—only 10° below body-heat—which I, therefore, assume to be the natural temperature of the breath, and which, in the action of breathing, is totally dissipated and lost in the open air.

My invention aims at economizing and utilizing this wasted heat by any simple contrivance for conveying it to our feet, where it is so much needed.

I have found, by actual experience, that the tubes [in Figs. 1 and 2] in a short time become warmed by the body, so that little heat of the breath is lost in its passage to the feet; that, accordingly, the air I find is delivered in boot or shoe with a temperature of about 84° Fahrenheit—a loss of only 4°.

After a few sharp blasts of breath at the beginning—which may be repeated at intervals—it becomes only necessary to inhale naturally with closed, and exhale with open, lips—an easy process, which I have ascertained practically may be kept up a long time, as, for example, for miles on a railroad-car, without much personal inconvenience.

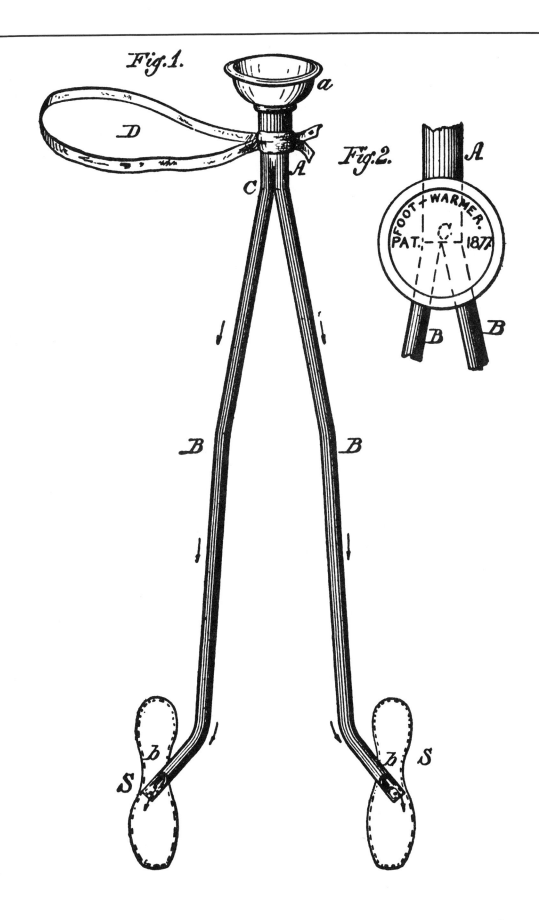

Luminous Hat
Patent No. 273,074 (1883)
Robert F. S. Heath of Camden, New Jersey

In Fig. 1 *A* represents a hat of well-known form, the surface of which is self-luminous, which may be produced by dipping the hat into "self-luminous" material, painting a hat with such material, forming a sheathing of the material, and fitting it over the hat, molding the hat of said material, either by itself or in combination with another material or other materials, or in any other suitable mode, and the body, brim, band, &c., of the hat may be made entirely or partly self-luminous, it being evident that the invention is equally applicable to caps, bonnets, and all articles of head-wear, regardless of the material, style, shape, and ornamentation thereof, or of the mode of making the same.

Among the advantages of the invention are the facility of seeing and finding the hat, &c., in closets and dark rooms and places, the presentation of a hat, &c., of different shades during day and night, the beautiful appearance of the article when worn at night, and the provision of distinguishing or indicating the localities of those who may wear the hats, &c., whose occupations are dangerous, such as miners, mariners, &c. For persons who are exposed to weather, sea, &c., the head-wear will be suitably waterproofed, so that the self-luminous nature thereof will not be injured by water.

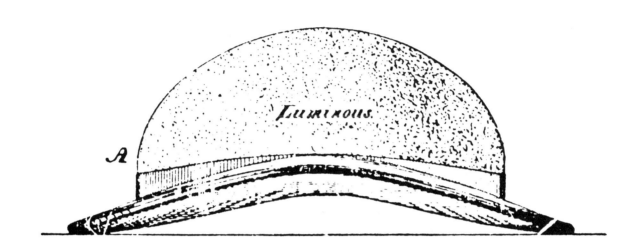

Raincoat with Drain
Patent No. 273,115 (1883)
John Maguire of Toronto, Canada

The object of the invention is to provide a water-proof coat which can be worn in rainy weather without the wearer's legs being made wet from water dripping off the skirt of the coat; and it consists of a water-proof coat having the bottom edge of its skirt turned up, forming a trough or channel to receive the water flowing on the surface of the coat, suitable provision being made to carry off the water away from the legs of the wearer of the garment.

Figure 1 is a view of my improved water-proof coat as it will appear when in use. Fig. 2 is an enlarged detail, showing the form of the trough made in the bottom edge of the skirt. Fig. 3 is an enlarged sectional detail, showing the arrangement of the spout for carrying off the water from the trough.

In the drawings, A represents the skirt of a water-proof coat, having its bottom edge turned up to form a trough, B.

C is a piece of rubber or other suitable material, placed within the trough to form its bottom and to keep the edge turned up away from the side of the coat. Tie-pieces D are placed around the edge for the same purpose, and in order to hold the edge in its upright position.

E is a spout or tube leading from the trough, and designed to carry away the water as it accumulates in the trough. It will be seen that this tube carries the water away from the legs of the wearer of the garment, and may be made any length or shape and placed at any convenient point or points in the bottom edge of the skirt. It will also be seen that a coat provided with a trough of this description completely protects the wearer, the water on the surface of the coat being carried away from his legs.

Although the coat is specially designed for gentlemen's use, it will of course be understood that ladies' coats may be similarly made.

Fig. 1.

Fig. 2. *Fig. 3.*

Clothes Brush Flask

Patent No. 490,964 (1893)
Thomas W. Helm of Danville, Virginia

Fitting snugly to the base of the contracted neck a of the flask is a cup D which, when applied to the neck of the flask as shown in Fig. 1, presents an outer surface flush with that of the flask and fitting snugly against the brush block, this cup conforming to the shape of that part of the block A which it covers, so that when the cup is in place on the flask the brush presents substantially the same appearance as an ordinary clothes brush, the cup, however, being readily removed when it is desired to use the same for drinking purposes, as shown in Fig. 2.

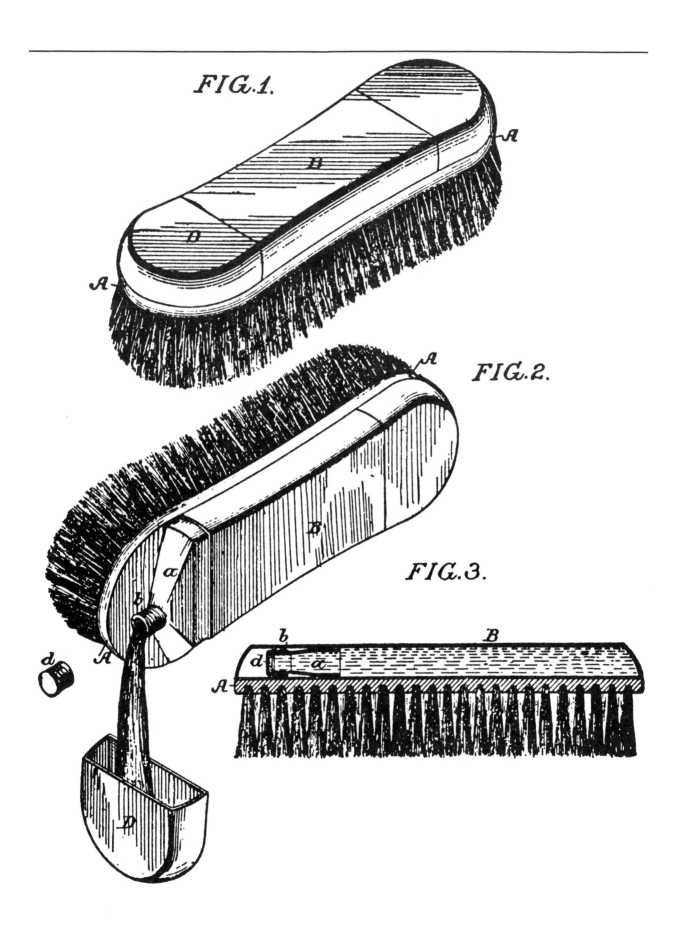

Zipper
Patent No. 504,038 (1893)
Whitcomb L. Judson of Chicago, Illinois

My invention relates to clasp lockers or unlockers for automatically engaging or disengaging an entire series of clasps by a single continuous movement.

The invention [Figs. 1–10] was especially designed for use as a shoe-fastener; but is capable of general application wherever clasps consisting of interlocking parts may be applied, as for example, to mail-bags, belts, and the closing of seams uniting flexible bodies. To these ends, the clasps are made with interlocking parts, which, when in position, can only engage with each other when at an angle to the line of strain. The clasps have underreaching and overlapping projections or lips at their forward ends which prevent the engagement or disengagement of the hook-portions of the clasps, except when thrown upward, so that the parts stand at an angle to each other of about ninety degrees. These clasps or fasteners, when in position on the flaps of a shoe or other adjacent parts which are to be united, may be engaged one at a time in succession, by bringing the two parts of the clasp into their proper angular relation to each other, by hand. But this is a tedious operation; and makes it difficult to draw the adjacent parts together, under the proper strain. I therefore provide a hand device, consisting of a movable guide, having cam-ways for permitting the passage of the clasps, by the movement of the guide from one end to the other of the series; and the cam-ways are so shaped and related that by the passage of the guide in the other direction, the clasps will be disengaged and separated. In other words, one end of the guide has two channels or grooves, for receiving the parts of the fasteners when open or disengaged, and this may be called the forward end of the guide. The other or back end of the guide has a single channel or cam-way, into which the two channels from the forward end converge over an angular center ridge or instep. By moving the guide, so that the separate parts of the clasp enter the respective channels or cam-ways at the front of the guide, the entire series of clasps will be delivered from the other or rear end of the guide properly engaged together.

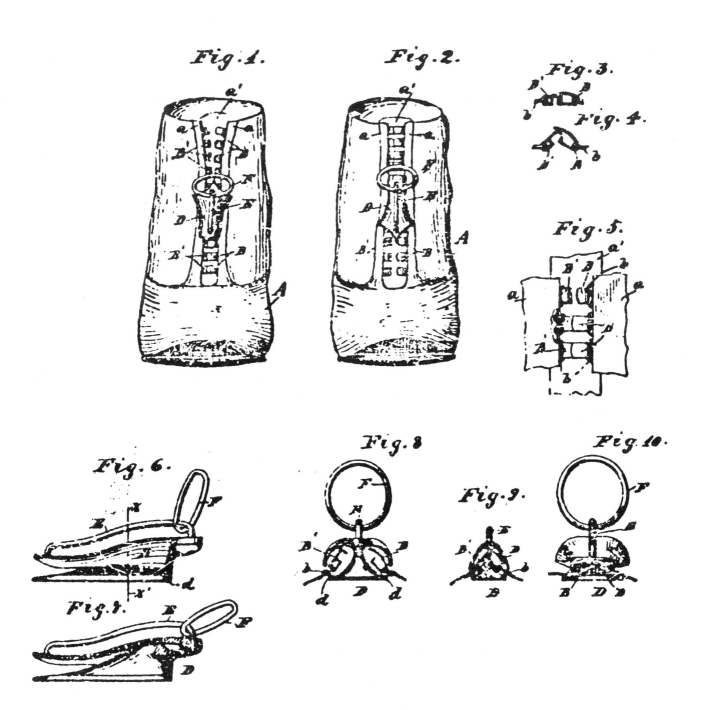

Saluting Hat
Patent No. 556,248 (1896)
James C. Boyle of Spokane, Washington

This invention relates to a novel device for automatically effecting polite salutations by the elevation and rotation of the hat on the head of the saluting party when said person bows to the person or persons saluted, the actuation of the hat being produced by a mechanism therein and without the use of the hands in any manner.

Should the wearer of the hat having the novel mechanism within it desire to salute another party, it will only be necessary for him to bow his head to cause the weight-block 30 [Figs. 1 and 2] to swing forwardly [which] will, by the consequent vibration rearwardly of the upper end of the arm 29^a, push the rod 31 backward and release the stud 34 on the rock-arm 32 from an engagement with the lifting-arm 27, so that the latter will, by stress of the spring 16, be forcibly rocked down into contact with the pin 33, as indicated by dotted lines in Fig. 2, the arm 28 having been correspondingly moved toward the lift-pin f. When the person making a salutation with the improvement applied to his hat resumes an erect position after bowing, the weight 30 will swing back into a normal position, which will draw the upper end of the rock-arm 32 forwardly and move its lower end rearwardly far enough to release the arm 27 from the pin 33. The gear-wheel 25 will now be moved by the spring 16, so as to impinge the short arm 28 on the lower side of the stud f, which will cause the guide-plate 15 to slide upward, carrying the post 14 with it. Just before the arm 28 passes the stud f the detent-spring q will press its curved toe q through the slot in the front plate of the case 10 and project said toe below the rounded lower end of the post 14. The lifting-arm 27 is now brought into contact with the pin e, and the pressure of the said arm on the pin e causes the post 14 to move upwardly in the depression c of the guide-plate 15 until it enters the slot d. The lift-pin e will now be swung through the rear portion of the cross-slot d by the arm 27, and by the impetus given to the pin and post 14 by said arm the post, bow-piece, and hat A will receive a rotary movement sufficient to bring the pin e into the depression c, when the gravity of the parts will cause the hat to drop into its normal position on the wearer's head.

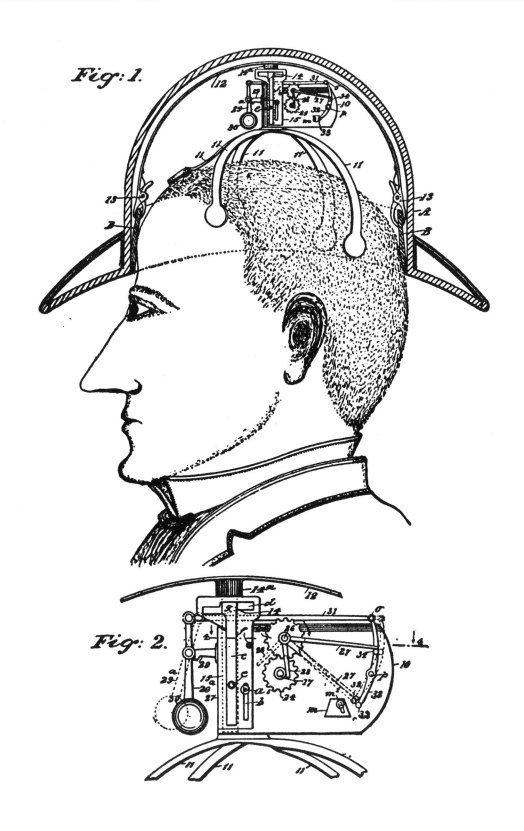

FIVE
Ladies' Liberation

Beyond the wondrous twentieth-century devices created with the woman's traditional household chores in mind—those that have made it easier for her to cook, clean, wash, and sew—there have been precious few inventions to ease the ancient burdens of her body.

In fact, with the exception of the women's liberation movement (which some persist in arguing is an invention of the mind), the only major patents designed to free a lady from things other than *kinder, kuchen,* and *kirchen* have been the brassiere in 1914 and the contraceptive pill in 1955. Even in those two cases, it was largely a matter of women helping themselves.

Mary Phelps Jacob (later known as Caresse Crosby) was a well-known New York socialite during the early 1900s. In those days fashionable women wore restrictive whalebone corsets and primitive bust supporters (see p. 98). One night, as Miss Jacob and her French maid, Marie, were preparing for a debutante ball, they discarded the bulky corset and—from two pocket handkerchiefs, some pink ribbon, and a piece of thread—devised a much lighter alternative. The following year Miss Jacob took out a patent on her "Backless Brassieres," but the few hundred samples she and Marie produced did not sell. Nevertheless

> "After all," as a pretty girl once said to me, "women are a sex by themselves, so to speak."
> *Max Beerbohm,*
> The Pervasion of Rouge

the Warner Brothers Corset Company bought the rights for $15,000; and the patent has since been estimated to be worth 100 times that much. Through it all Miss Jacob never forgot that she was a descendant of Robert Fulton. "I believe that my ardour for invention," she wrote in her memoirs, "springs from his loins. I can't say that the brassiere will ever take as great a place in history as the steamboat, but I did invent it, and perpetual motion has always been just around the corner."

Margaret Sanger established the world's first family-planning service in Brooklyn in 1914. Though it was closed down as a public nuisance and she was sent to prison, Miss Sanger would not give up her mission. In 1927 she organized the first World Population Conference; in 1948 she began the International Planned Parenthood Federation; and in 1951 she persuaded Dr. Gregory Pincus, an animal-fertility expert from Shrewsbury, Massachusetts, to develop an oral contraceptive. The Pill was patented four years later.

Most of the other devices shown herein were invented by men, in order that their ladies would either shape up (breast developer, dimple maker, nose shaper) or spin out (centrifugal delivery table).

Dimple Maker
No. 560,351 (1896)
Martin Goetze of Berlin, Germany

The present invention consists of a device which serves either to produce dimples on the human body or to nurture and maintain dimples already existing.

When it is desired to use the device for the production of dimples, the knob or pearl c of the arm a [in Fig. 1] must be set on the selected spot on the body, the extension d, together with the cylinder f, put in position, then while holding the knob n with one hand the brace i must be made to revolve on the axis. The cylinder f serves to mass and make the skin surrounding the spot where the dimple is to be produced malleable.

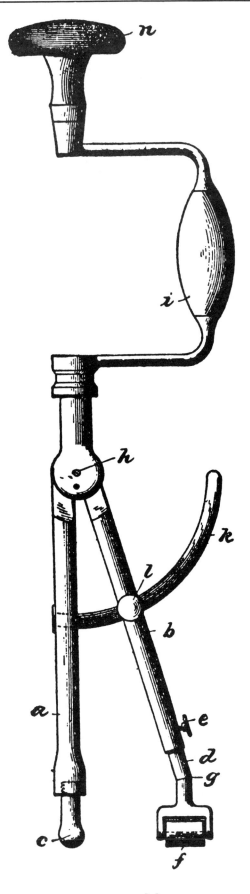

Bust Supporter
Patent No. 844,242 (1907)
Johannes Bree of Charlottenburg, Germany

This invention relates to bust-supporters, the object being to provide a bust-support which differs from the usual corset in that it can be worn without any injury to health.

It consists, essentially, of two back-plates or frame-pieces. These back-plates are connected together in any suitable manner, and to them the other essential parts of the bust-support are attached, so that when wearing the same neither the breasts nor the stomach nor the liver are tightened in by lacing.

Figure 1 shows the front view of a female form provided with one form of my improved bust-support. Fig. 2 is a corresponding back view. Fig. 3 shows a rear perspective view of a modified form. Fig. 4 shows the front view of a female form with a further form of bust-support applied.

As may be seen from Fig. 1, this support does not exercise any injurious pressure on any of the important organs of the body. From Fig. 2 it may further be seen that independently of this there is present a new feature in this bust-support, for by raising up, bending and stretching the arms, and by other movements of the same the essential parts of the bust-support act as an elastic system, the relative stable middle points of which are the back parts 1 and 2. These parts of the body have been specially chosen, so that by the arrangement and form of the back-pieces no injury can arise from pressure or friction of the shoulder-blades or from particular movements of the body or its extremities, since the spring connecting of the essential parts of the bust-support allows the latter to yield, extend, and be afterward drawn together again by such springs. The body-girdles 23 and 24, which are also in themselves not novel, exert a massage action on the body, owing to their being fastened to the springs 19-20 and 21-22, thus favorably assisting the work of the intestines. For increasing this action springs 27-28 or 29-30 (shown in Fig. 4) might be arranged, which elastically draws together the individual parts 24[a] and 24[b] or 23[a] and 23[b], from which the girdles are formed.

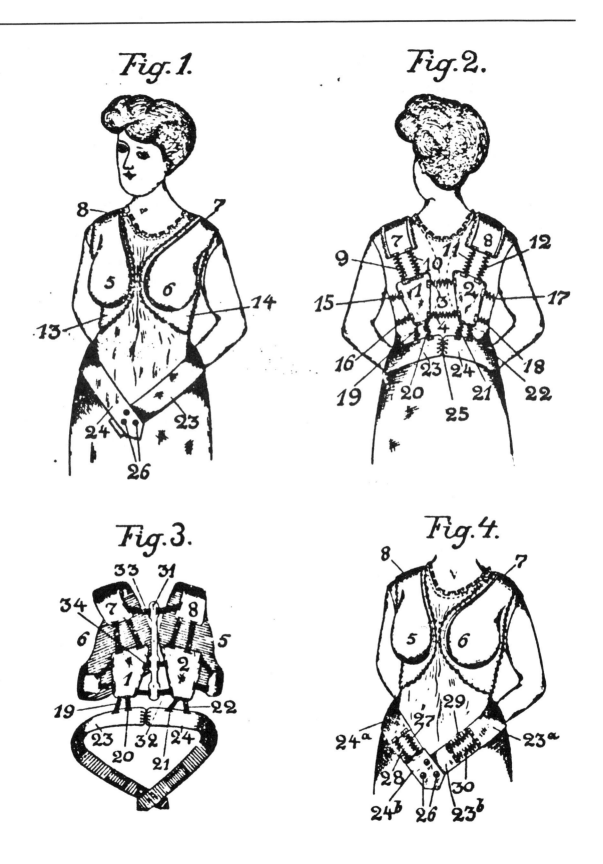

MOUSETRAPS AND MUFFLING CUPS

Nose Shaper
Patent No. 850,978 (1907)
Ignatius Nathaniel Soares of Framingham, Massachusetts

The noses of a great many persons are slightly deformed, and therefore because of the prominence of this feature the appearance of the face is more or less disfigured. Such deformity can frequently be remedied by a gentle but continuous pressure, and it is the object of this invention to bring about this result in a way that shall be painless to the individual.

Figure 1 represents a three-quarter view of a woman with the device in place when securing its functions. Fig. 2 is a side view showing the device partially in cross section. Fig. 3 is a plane view of the device.

The device is made up of a body portion A of a suitable resilient material capable of being bent into a desired shape and of retaining that shape, but in a slightly yielding manner, owing to the resilient quality of the material. A thin sheet metal properly tempered is best for this purpose. Since material of this sort would be unpleasant when placed upon the nose and would unless toughened not produce the necessary friction, a lining or casing B, of soft frictional material, such as chamois skin or fabric, is provided. It is only necessary that this lining should be on the side next to the nose; but it is preferable to make it in the form of a casing enclosing the material entirely. This may best be secured by cutting out two pieces of the shape shown in the outlines of Fig. 3, placing the resilient material A between the two pieces, and stitching the parts of the casing together on the lines indicated at $a\ b$.

The device is preferably provided with a projecting tongue C, which when bent into position rests on the bottom of the nose and aids in holding the device in place. The lining or casing should be cut to conform to this tongue, as indicated in the drawings. This tongue acts also to press up upon the cartilage between the nostrils when desired, and so to reshape the bottom of the nose.

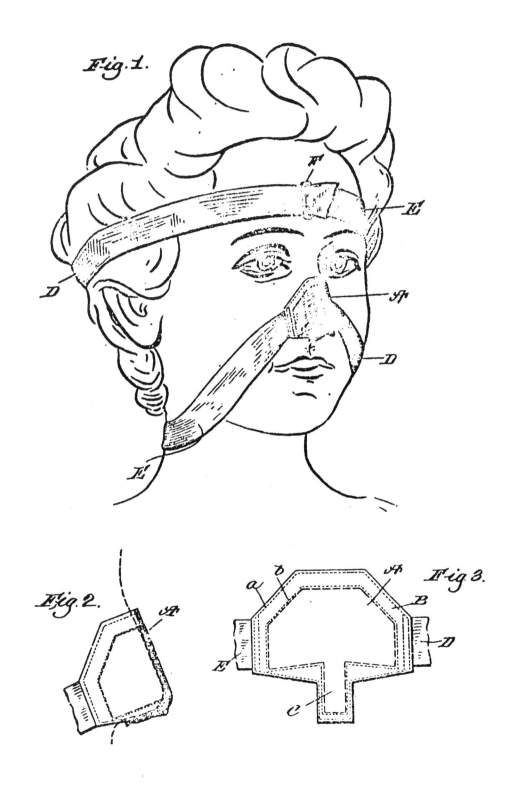

Modest Nurser
Patent No. 949,414 (1910)
Hugh B. Cunningham of Arnot, Pennsylvania

The primary object of this invention is an improved construction of a device for use by mothers with nursing infants, and designed particularly to avoid unpleasant and embarrassing situations in which mothers are sometimes placed in public places by the necessary exposure of the breast in suckling the child.

The invention consists essentially in a nursing attachment designed to be worn over the breasts and arranged for the detachable connection thereto of the nipple on a tube of any desired length, the nipple or nipples, according to whether there be one or two employed, being worn inside of the shirtwaist or other outer garment and it being only necessary when the child is to be nursed, to slip the nipple out of the waist, thereby avoiding the necessity of exposing the person.

In the practical use of the device [see Fig. 1], the shields 1 are adjusted over the breasts with the cups 4 directly over the nipples and the cap or caps 5 are then attached, the nursing nipples 7 and all other parts being hidden beneath the wearer's waist. Whenever the child requires nursing it is only necessary to slide one of the nursing nipples 7 [Fig. 2] out from the waist and the child can obtain its proper nourishment without the exposure of the mother's person and the consequent embarrassment which is thus often occasioned.

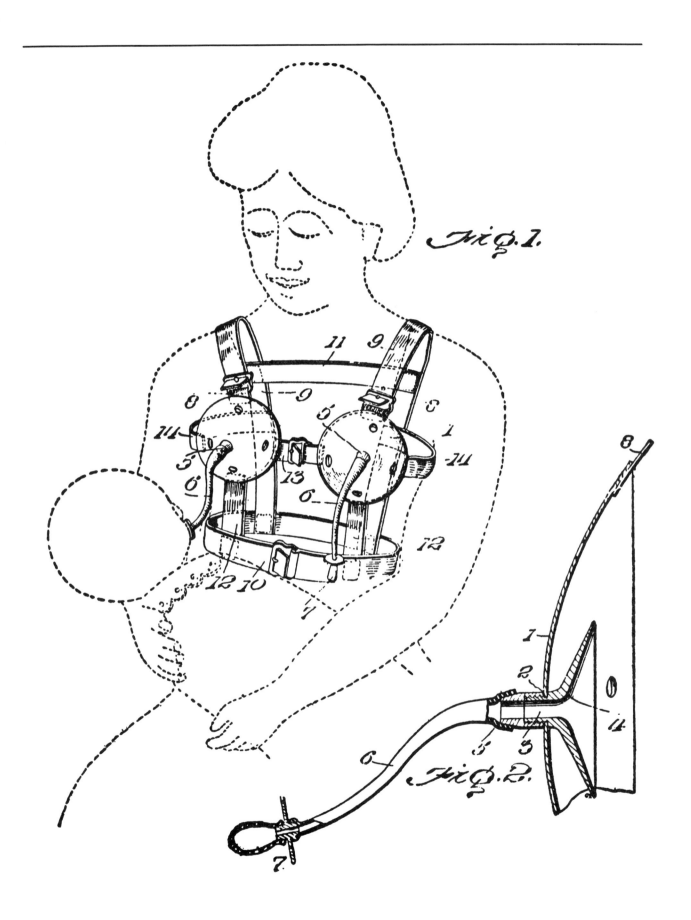

Responsive Brassiere
Patent No. 1,033,788 (1912)
Oliver C. Dennis of Chicago, Illinois

The invention designs to construct the bust-forms so that when worn, they will not have the appearance of lifeless members, but will vibrate responsively to movements of the wearer.

In use, when pneumatic pads are inflated and covered by garments, the bosom is sometimes rather lifeless in appearance, due to the inflated pads which do not readily or freely respond to the movements of the wearer's bosom. To overcome this appearance, a weight f [Figs. 1–3] is molded into the front of the casing C, and under the influence of the body is suspended at the front of the pad and sets the front in motion so that it will vibrate freely in response to any movement of the body of the wearer.

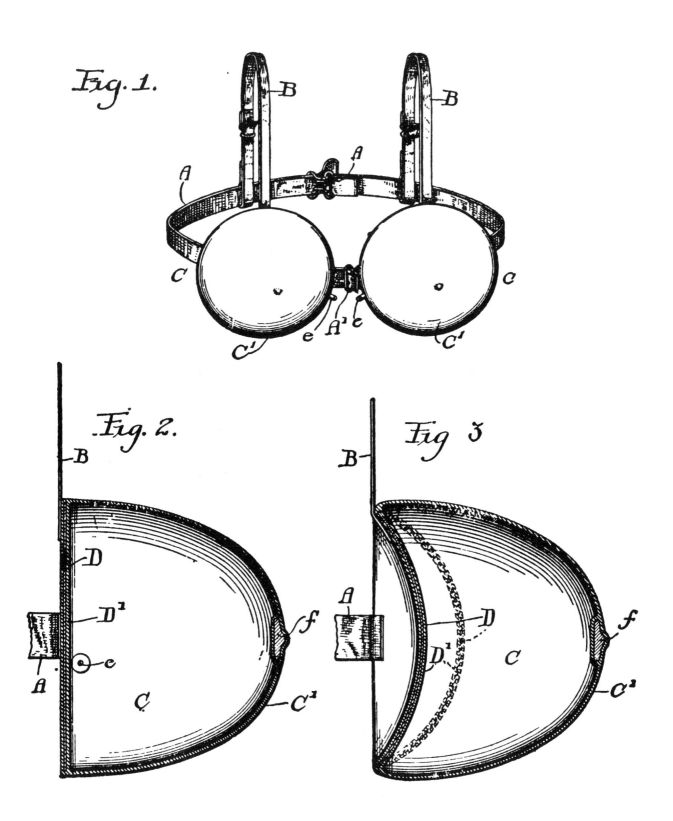

Massage Apparatus
Patent No. 1,175,513 (1916)
Louis W. G. Flynt of Chicago, Illinois

Mechanical massage heretofore has been impracticable for the reason that where a tool has been used, it was almost impossible for the user to apply the tool to all parts of the body. This I accomplish mechanically by constructing a tool adapted to surround the body of the user and which is expansible, whereby the tool will conform to the contour of the body of the user in its reciprocating movement.

The tool member shown in Fig. 2 may be made as large or as small as desired, according to the number of tool units that are coupled together. Where the body is to be massaged enough units are coupled together to surround the body of the user at its smallest circumference, for instance, at the waist. The tool may be opened by disconnecting the caps 16 and 17 so that it may be conveniently fixed in position around the body of the user whereupon the caps 16 and 17 are connected and the user thereupon grasps the handles 39 and 40. By moving the arms downwardly, the tool is pulled down along the body of the user, against the actions of the springs 30 and 31, until it reaches approximately the knees of the user, as shown in the diagrammatic position in Fig. 1.

In passing over the body of the operator, the tool expands or contracts, thereby automatically conforming to the exterior surface of the body, and as each unit comprising the tool is independently rotatable, and furthermore is capable of rotating on its own axis, the action of the tool simulates the rubbing or kneading action of a hand massage by the distorting and twisting action of the individual tool units in their rotating movement over uneven surfaces.

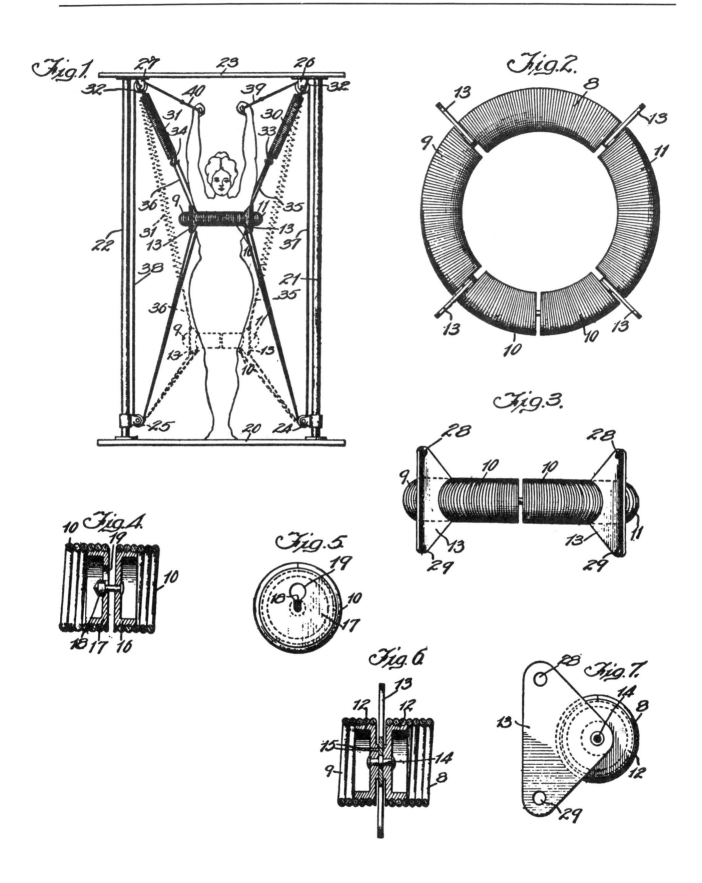

MOUSETRAPS AND MUFFLING CUPS 107

Gentleman's Cigarette Holder
Patent No. 1,488,412 (1924)
Allessandro Taraborrelli of Providence, Rhode Island

The principal object of the invention is to provide clamping means for clamping the cigarette to hold the same in place.

Another object of the invention is to make the device in the shape of a human being, using the arms as the clamping jaws and the legs the means for working the jaws.

In [Figs. 1–3] 1 indicates a tube having a mouth-piece 2 formed at one end and its other end adapted to receive the cigarette. On top of this tube is located the figure 3 which is formed of two halves by dividing the figure longitudinally. A pin 6 is secured to the tube and each half of the figure is provided with a perforated projecting part 5 for engaging said pin so that the two parts of the figure are pivotally supported upon the tube. A spring 8 is carried by the pin 6 and has its extremities engaging the legs 7 of the figure, this spring tending to press the upper parts of the two halves of the figure together with the arms 4 engaging the cigarette. When it is desired to remove the cigarette from the holder it is simply necessary to press inwardly upon the legs 7 to separate the arms of the figure to cause them to release the cigarette.

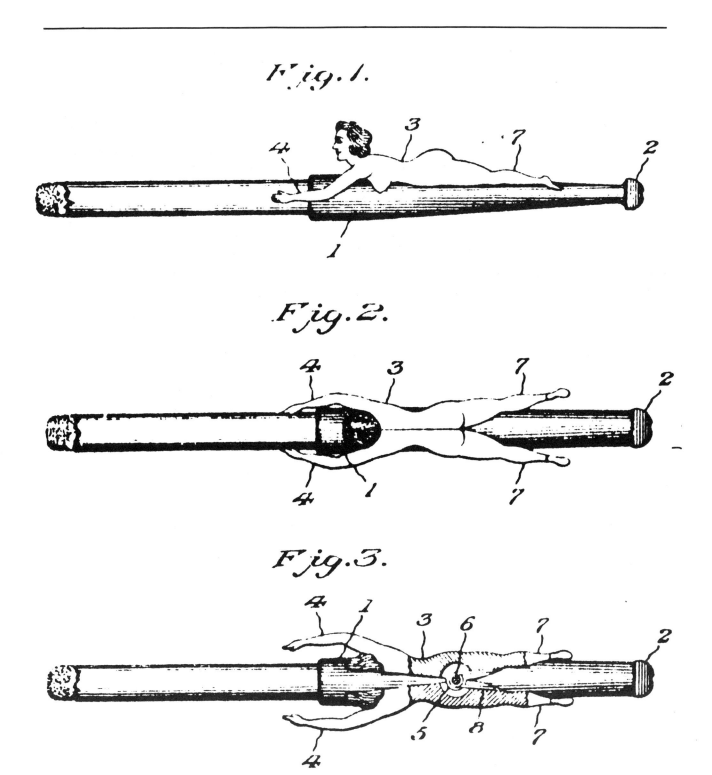

Lip Shaper
Patent No. 1,497,342 (1924)
Hazel Mann Montealegre of Iola, Kansas

This invention relates to devices for re-shaping the upper lip of a person, and has for its object the provision of a simple and easily applied device to re-shape the upper lip of a person to conform to what is known as the "Cupid's bow," whereby it is unnecessary to resort to a surgical operation to produce this effect.

By my new and improved device, I not only cause a depression to be formed in the upper surface and centrally of the upper lip, but the upper lip will eventually be changed to the form of the well-known Cupid's bow. [See Figs. 1–5.]

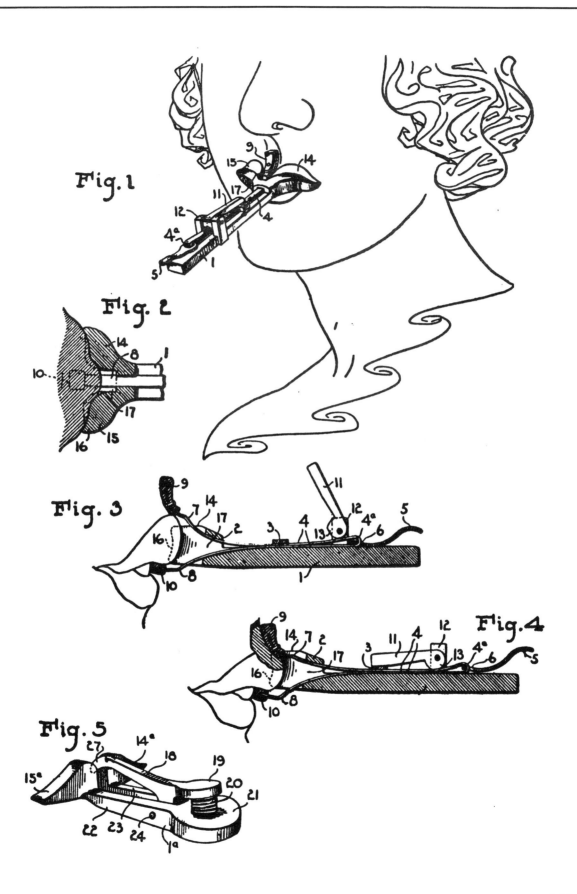

Centrifugal Delivery Table
Patent No. 3,216,423 (1965)
George and Charlotte Blonsky of New York, New York

It is known that due to natural anatomical conditions, the fetus needs the application of considerable propelling force to enable it to push aside the constricting vaginal walls, to overcome the friction of the uteral and vaginal surfaces and to counteract the atmospheric pressure opposing the emergence of the child. In the case of a woman who has a fully developed muscular system and has had ample physical exertion all through the pregnancy, as is common with all more primitive peoples, nature provides all the necessary equipment and power to have a normal and quick delivery. This is not the case, however, with more civilized women who often do not have the opportunity to develop the muscles needed in confinement.

It is the primary purpose of the present invention to provide an apparatus which will assist the under-equipped woman by creating a gentle, evenly distributed, properly directed, precision-controlled force, that acts in unison with and supplements her own efforts.

In accordance with the invention, there is provided rotatable apparatus capable of subjecting the mother and the fetus to a centrifugal force directed to assist and supplement the efforts of the mother so that such centrifugal force and her efforts act in concert to overcome the action of resisting force and facilitate the delivery of the child.

Fig. 4 is a side elevational view of a modified form of the means for supporting the mother on the apparatus; and

Fig. 5 is a view similar to Fig. 4 showing another embodiment of such means.

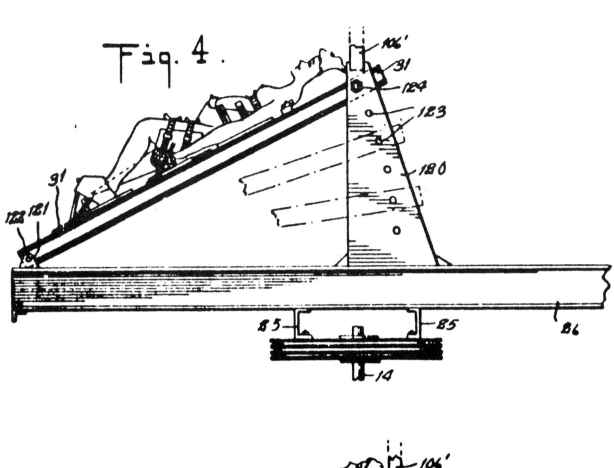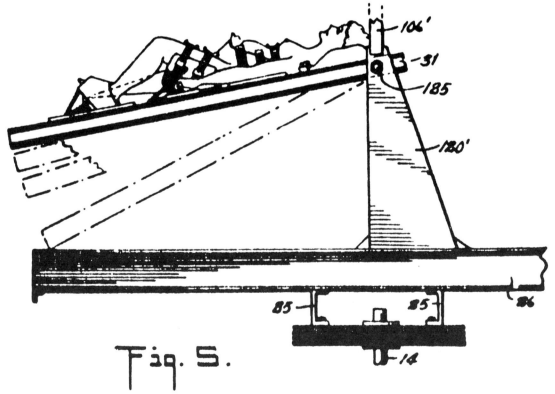

Breast Developer
Patent No. 3,913,569 (1975)
Thomas S. Kanonas of Windsor Locks, Connecticut

This invention [see Fig. 1] consists of a set of identical plungers each having a handle and a resilient hemi-spherical cup member with a foam lining disposed completely about the interior surface thereof, the apparatus adapted to have the cups placed over a woman's breasts and, through manipulation thereof by use of the handles in a back and forth direction, create suction with the body surrounding the breast to draw the breast into the cup to stimulate the breast and to enhance and develop the size of the breast through frequent exercise thereof.

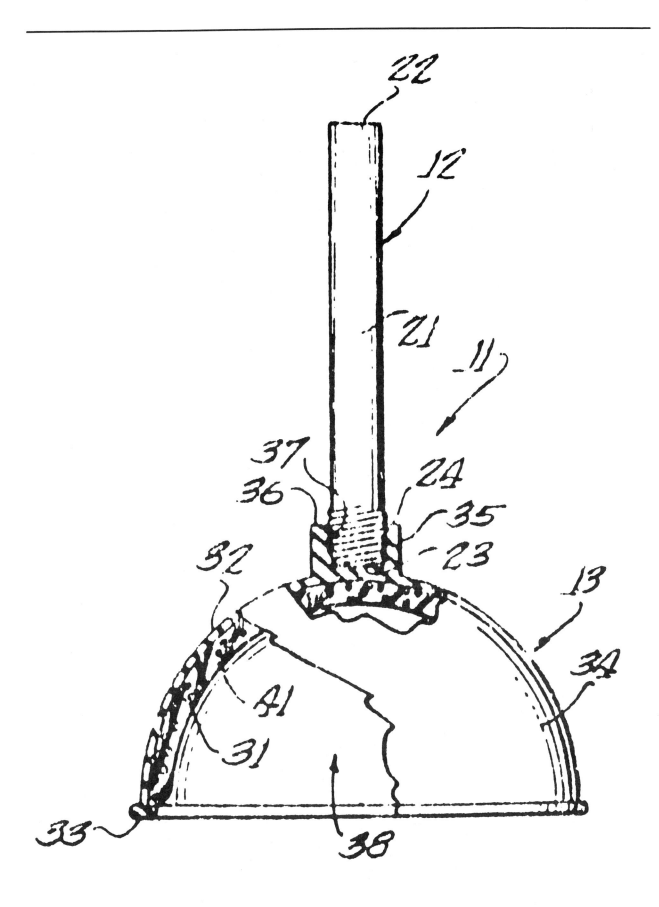

SIX

Way Down (and Way Out) on the Farm

The Industrial Revolution was already under way by the time Eli Whitney happened onto the cotton fields outside Savannah in 1793. Much of it was brought about by landmark inventions in the making of cloth. In 1738 John Kay introduced the flying shuttle, in 1764 James Hargreaves the spinning jenny, in 1769 Richard Arkwright the water frame, in 1777 Samuel Crompton the spinning mule, and in 1787 Edmund Cartwright the self-acting loom. But at that time over ninety percent of all clothing was made from wool or flax—because, though cotton was light and plentiful, it took one worker a full day to pull the seeds out of fifty pounds of fiber.

Whitney was born in 1765 in Westboro, Massachusetts. He had always been interested in what made things work; at age twelve he took apart and reassembled his father's watch, about the same time made his own violin, and he worked his way through Yale by doing mechanical services. At twenty-eight he went to Savannah, Georgia, where he met "a number of very respectable Gentlemen who all agreed that if a machine could be invented which would clean cotton with expedition, it would be a great thing both to the Country and to the inventor." This is the way he told the rest of the story, in a letter to his father:

> In about ten Days I made a little model, for which I was offered, if I would give up all right and title to it, a Hundred Guineas. I concluded to relinquish my school and turn my attention to perfecting the Machine. I made one before I came away which required the labor of one man to turn it and with which one man will clean ten times as much cotton as he can in any other way before known and also cleanse it much better than in the usual mode. This machine may be turned by water or with a horse, with the greatest ease, and one man and a horse will do more than fifty men with the old machines. It makes the labor fifty times less, without throwing any class of People out of business.
>
> I returned to the Northward for the purpose of having a machine made on a large scale and obtaining a Patent for the

> We know that the tail must wag the dog,
> for the horse is drawn by the cart;
> But the Devil whoops, as he whooped of old;
> "It's clever, but is it Art?"
> *Rudyard Kipling,*
> The Conundrum of the Workshops

> invention. I went to Philadelphia soon after I arrived, made myself acquainted with the steps necessary to obtain a Patent, took several of the steps and the Secretary of State Mr. Jefferson agreed to send the Patent to me as soon as it could be made out.
>
> How advantageous this business will eventually prove to me, I cannot say. It is generally said by those who know anything about it, that I shall make a Fortune by it. I have no expectation that I shall make an independent fortune by it, but think I had better pursue it than any other business into which I can enter. Something which cannot be foreseen may frustrate my expectations and defeat my Plan; but I am now so sure of success that ten thousand dollars, if I saw the money counted out to me, would not tempt me to give up my right and relinquish the object. I wish you, sir, not to show this letter nor communicate anything of its contents to any body except My Brothers and Sister, enjoining it on them to keep the whole a profound secret.

Alas, the cotton gin (p. 118) was not a secret for long. Whitney proved to be a much better inventor than businessman; he ran into all kinds of difficulties in protecting his hard-earned patent, and didn't make nearly the fortune from it that he had sought and expected.

But the gin, just as it helped fuel the Industrial Revolution, inaugurated an agricultural one. The first half of the nineteenth century saw a number of improved reapers, the most successful of which was patented by Cyrus McCormick of Staunton, Virginia, in 1834 (p. 120). By the 1870s various attachments had been invented so that the entire process of shearing, baling, and binding could be done by one machine.

Taking care of the chickens and the cows, though, was and remains another story—particularly on the thousands of family farms where the chickens are still fed table scraps, the cows are still milked by hand, and somebody's still trying to figure out a better way to do the chores.

Cotton Gin
Patented in 1794
Eli Whitney of Westboro, Massachusetts

The principal parts of this machine are 1st, The frame, 2d, The cylinder, 3d, The breastwork, 4th, The clearer, & 5th, The hopper.

The frame, by which the whole work is supported and kept together, is of a square or parallelogramic form and proportioned to the other parts as may be most convenient.

The cylinder is of wood, its form is perfectly described by its name, and its dimensions may be from six to nine inches in diameter, and from two to five feet in length. This cylinder is placed horizontally across the frame, leaving room for the clearer on one side, and the hopper on the other. In the cylinder is fixed an iron axis which may pass quite through, or consist only of gudgeons driven into each end.

The breastwork is fixed above the cylinder, parallel and contiguous to the same. It has transverse grooves or openings through which the rows of teeth pass as the cylinder revolves and its use is to obstruct the seeds while the cotton is carried forward through the grooves by the teeth. The thickness of the breastwork is two and half or three inches, and the under side of it is made of Iron or brass.

The clearer is placed horizontal with and parallel to the cylinder. Its length is the same as that of the cylinder, and its diameter is proportioned by convenience. There are two, four or more brushes or rows of bristles, fixed in the surface of the clearer in such a manner that the ends of the bristles will sweep the surface of the cylinder.

One side of the hopper is formed by the breastwork, the two ends by the frame, and the other side is moveable from and towards the breastwork, so as to make the hopper more or less capacious.

The cotton is put into the hopper, carried thro' the breastwork by the teeth, brushed off from the teeth by the clearer and flies off from the clearer, with the assistance of the air, by its own centrifugal force. The machine is turned by water, horses or in any other way as is most convenient. [See Figs. 1–3.]

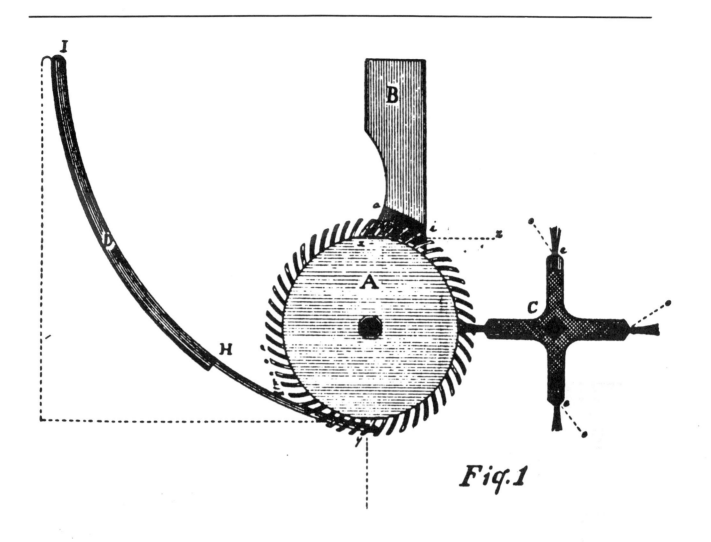

Fig. 1

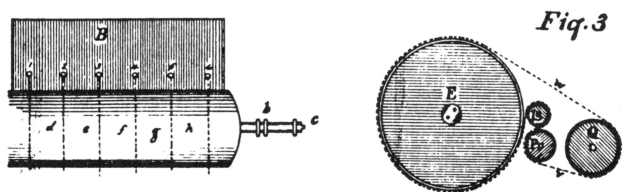

Fig. 2

Fig. 3

Combine Harvester
Patented in 1834
Cyrus H. McCormick of Rockbridge County, Virginia

Upon a frame of wood is to be constructed a platform of about six feet in width by about four or five in length [see Fig. 1]. . . . From the end of [the] broad piece nearest the platform rises a circular brace, secured to the reel-post by a movable screw-bolt, to allow for advancing or drawing back as the adjustment of the cutting may require. About three-quarters of a foot on the other end is a movable screw-bolt passing through both pieces, also allowing for a rise or fall in adjusting the height of cutting. . . . At a short distance in front are to be secured an arm on each side projecting toward the middle, where they are united and serve to throw the stalks of the grain toward the cutting apparatus. . . . Below the inner shaft from the single-tree end is secured a long bow continuing along the direction of the shaft to the front of the horse, where it passes round and joins to the other shaft. The object of this bow is to throw the stalks inward toward the cutting apparatus. . . . [The] end of the platform is to be closed by a strip of cloth stretched along it, and as high as the stalks. . . . On the axis hung between the head-pieces is a wheel of about two-feet diameter, having the circumference armed with teeth to hold to the ground by. On the right of this wheel is another of about thirteen inches diameter, or containing thirty teeth and on the right side are to be the teeth in a smaller cog-wheel of about three and a half inches in diameter, or nine cogs sloping back toward the front of the platform, where is secured another cog-wheel of about eleven inches in diameter, containing twenty seven teeth. . . . To the lower of these cranks is attached, a long cutter of steel, notched on its lower edge like a reaping-hook with the grooves running in a line toward the right of the machine. This blade is attached to the frame-piece so that the motion is described in part of a circle. This motion, when the stalks are presented, cuts them through. Above this cutter slides another long plate to the upper crank of the same length, and secured in the same manner. . . . [These teeth] are to project in a line sloped in an opposite direction to the grooves in the cutters below, and their motion in sliding backward and forward is also contrary, thereby collecting the stalks as they come in contact with these teeth and force them across the teeth in the cutter below, thereby greatly assisting in the act of cutting them through. . . .

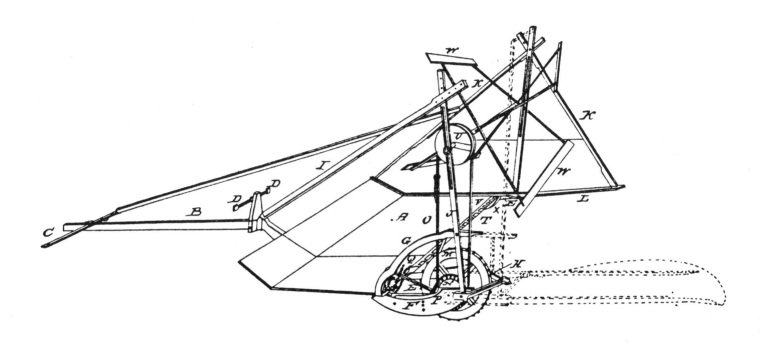

Barbed Wire
Patent No. 157,124 (1874)
Joseph F. Glidden of De Kalb, Illinois

This invention has relation to means for preventing cattle from breaking through the wire fences; and it consists in combining, with the twisted fence wires, a short transverse wire, coiled or bent at its central portion about one of the wire strands of the twist, with its free ends projecting in opposite directions, the other wire strand serving to bind the spur-wire firmly to its place, and in position, with its spur ends perpendicular to the direction of the fence-wire, lateral movement, as well as vibration, being prevented. It also consists in the construction and novel arrangement, in connection with such a twisted fence-wire, and its spur-wires, connected and arranged as above described, of a twisting-key or head-piece passing through the fence-post, carrying the ends of the fence-wires, and serving, when the spurs become loose, to tighten the twist of the wires, and thus render them rigid and firm in position. [See Figs. 1–3.]

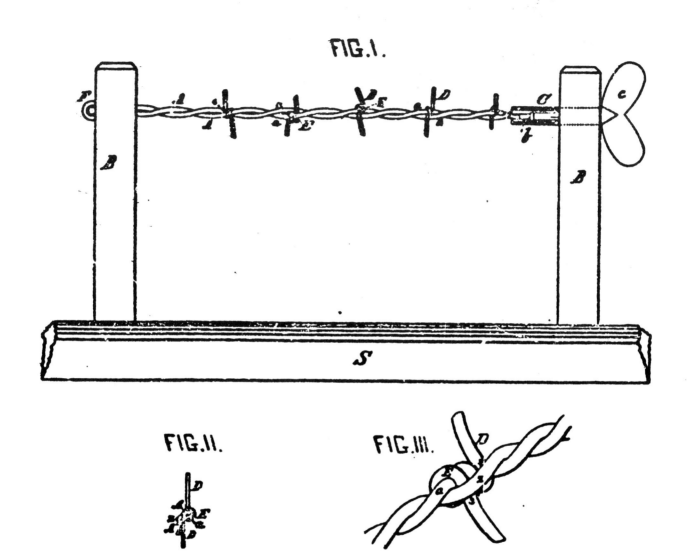

Cow Tail Holder

Patent No. 255,233 (1882)
Thomas Andrus and Napoleon Prior, both of North Wolcott, Vermont

This invention is an improved device for supporting a cow's tail during the operation of milking; it has for its object to provide a convenient means whereby, while the animal is allowed a certain amount of liberty to move her tail, it will be impossible to switch it into the milk-pail, or switch it in the face of the milker. These objects we attain by the devices illustrated in the accompanying drawings, in which Figure 1 represents a perspective view of our device entire, showing the method of its use, and Fig. 2 a side elevation.

In using our improvement the members of the spring-hook are pressed apart and the cow's tail at the hairy portion is inserted between the rectangular interlocking hooks at the extremities of the member of the spring-hoof, by which it is clasped and held, as shown.

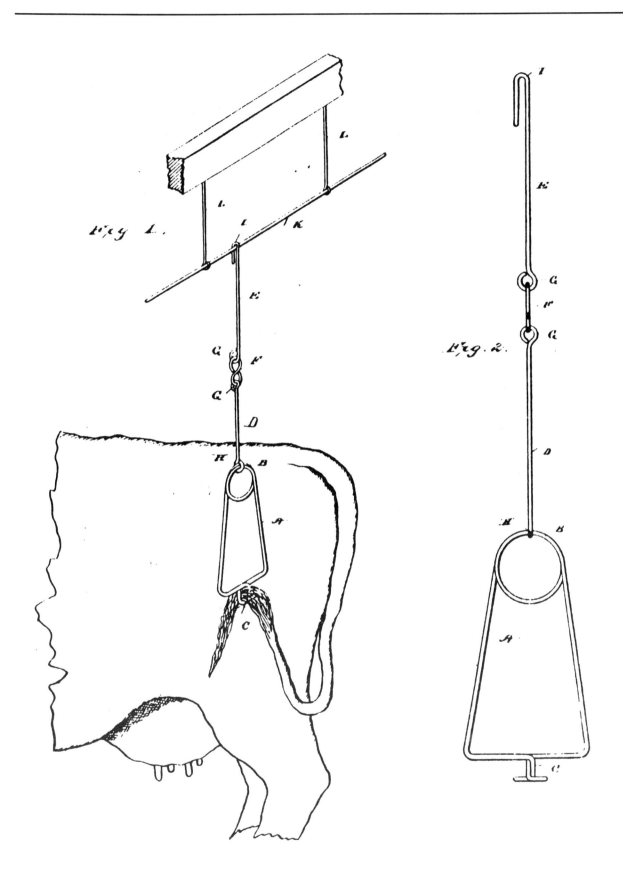

Milking Stool
Patent No. 359,921 (1887)
Allen B. Cowan of Hall's Valley, Ohio

In operation, the wearer buckles the waist-strap around his or her waist, for this stool is peculiarly adapted for use by women, the stool hanging down behind out of the way, as shown in Fig. 1 of the drawings, leaving both hands free to carry two pails. As soon as the wearer is ready to sit down to milk, by merely leaning slightly forward, as one sits, the stool swings directly underneath the person, and one can sit down upon it without touching it with the hand. Then, if the cow should move away a few feet or commence to kick, the person milking can get up quickly, and catch up the buckets with both hands without paying any attention to the stool, and follow up the cow, sitting down as before.

The object of swiveling the lower end of the waist board to the seat of the stool [Figs. 2 and 3] is that, in going into a narrow stall to milk, the wearer can walk in and sit down sidewise to the cow, where there is not room to turn and sit down facing the cow in the first place, and then turn on the stool, so as to face the cow, while the stool itself remains in its first position, the waist board turning on its pivot to allow of this movement.

Hen Goggles
Patent No. 730,918 (1903)
Andrew Jackson, Jr., of Munich, Tennessee

This invention relates to eye-protectors, and more particularly to eye-protectors designed for fowls, so that they may be protected from other fowls that might attempt to peck them, a further object of the invention being to provide a construction which may be easily and quickly applied and removed and which will not interfere with the sight of the fowl.

An additional object of the invention is to provide construction which may be adjusted so that it will fit different-sized fowls.

Figure 1 is a view showing the device attached to the head of a chicken. Fig. 2 is a perspective view of the device removed from the head of the chicken.

In the frames 5 and 6 are secured sheets 8 of glass, mica, or other suitable material, these sheets being concavo-convex, the concave sides being disposed toward each other, while the convex sides are disposed outwardly. The frames 5 and 6 are somewhat larger than the eyes of the fowl to be protected, and to adjust the frames they are drawn apart and sprung over the head of the chicken, so that one frame encircles each eye.

Hen Exerciser
Patent No. 828,227 (1906)
William Jared Manly of Erie, Pennsylvania

The main object of the present invention is the production of a device so constructed and arranged that the fowl in order to secure the food carried by the stretcher is compelled to undergo increased exercise as compared with the ordinary manner of feeding, whereby the fowls in the act of feeding are given that degree of exercise best suited to insure their proper condition.

In operation the fowls desiring the food in the box and stepping upon the platform cause the same to revolve under their weight, with the result that they are compelled to move rapidly forward to maintain such position on the platform as will enable them to reach the food in the box. It is of course to be understood that in all relative positions of the parts the supporting-plate 5, and therefore the platform, is disposed at an angle to the base 1, whereby to provide for the necessary movement of the platform under the weight of the fowl.

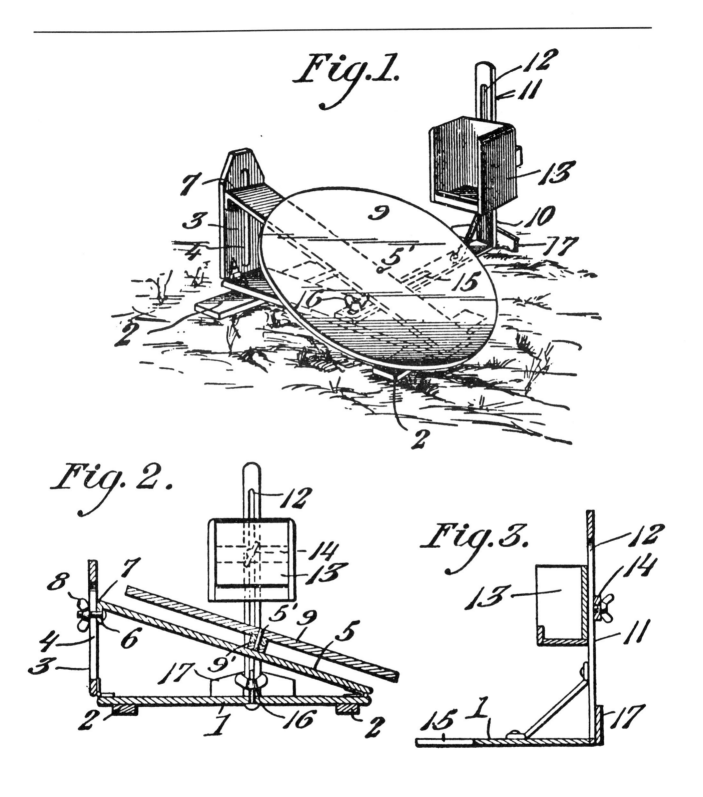

Egg Marker
Patent No. 970,074 (1910)
Stanley A. Merkley of Buffalo, New York

The primary object of my invention is the provision of a marking-device of this character which can be easily attached to the vent of a hen so that it will always be in place for marking an egg laid by said hen.

A further object of my invention is to so construct the marking-device that it will yield with the wall of the vent as the egg is being laid, thus permitting the egg to pass through the marking-device.

Figure 1 is a perspective view of a complete marking-device ready to be applied to a hen. Fig. 2 is a view of an egg showing the manner in which it is marked by the device shown in Fig. 1. Fig. 3 is a perspective view of the marking-device showing a different arrangement of marking-elements and a modified form of supporting-ring. Fig. 4 is a view of an egg showing the manner in which it is marked by the marking-device illustrated in Fig. 3. Fig. 5. is a view showing the manner in which the marking-device is attached to a hen, the device being in normal position. Fig. 6 is a similar view showing the marking-device expanded and an egg passing therethrough in its passage from the vent of the hen. Fig. 7 is a cross-section through one of the marking-elements. Fig. 8 is a view of a modified form of the securing-ring. Fig. 9 is a view of a marking-device provided with a shield or protector into which the marking-element is to be drawn when the device moves into normal position, as shown in dotted lines in said figure. Fig. 10 is an enlarged longitudinal section of a portion of the device showing the marking-element within the shield or protector.

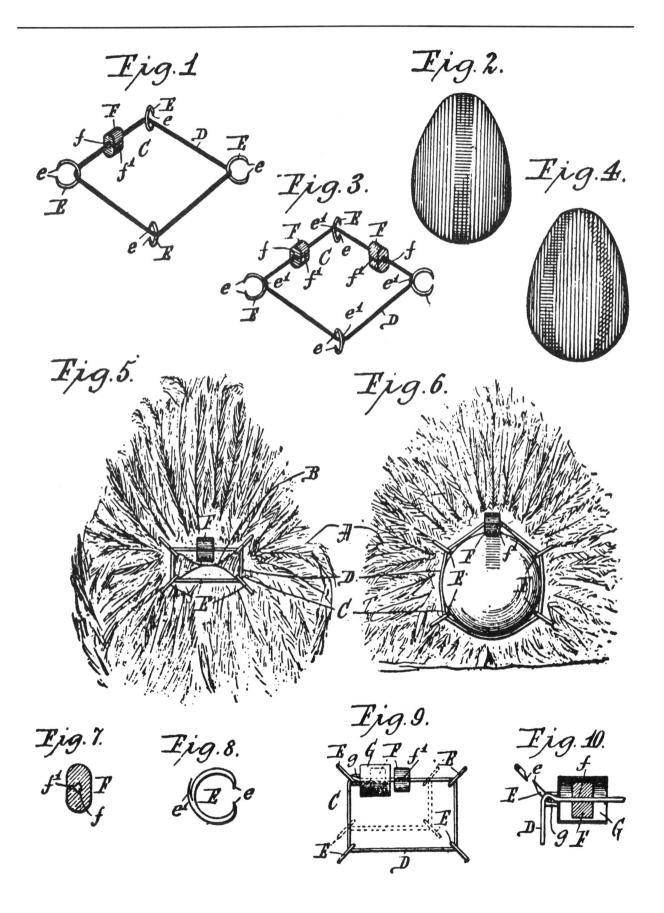

Sheepdog Nose Hook
Patent No. 1,046,177 (1912)
William Graham of Cookshire, Quebec, Canada

My invention relates to devices to prevent dogs from worrying sheep, and it has for its object [a device] which may be fastened to the nose of a dog [see Figs. 1–3], and which is provided with hooks which will become entangled in the wool of a sheep so that when the sheep starts to run the dog's nose will be pulled, and the dog will receive a lesson which will break him of his habit of worrying sheep.

When the hooks 19, with the stud 18, are secured to the ring member 5, which is attached to the nose of a dog, the hooks will become entangled in the wool of any sheep which the dog may attempt to worry. As soon as the dog is in close enough contact which the sheep to permit the hooks 19 to become entangled in the wool of the sheep, the sheep will start to run, which will yank the ring member 5, and give the dog's nose a severe pull, and, after a few attempts have been made in this way to worry the sheep, it will be found that the dog is very careful not to go too close to the sheep and that the sheep are no longer disturbed at the presence of the dog.

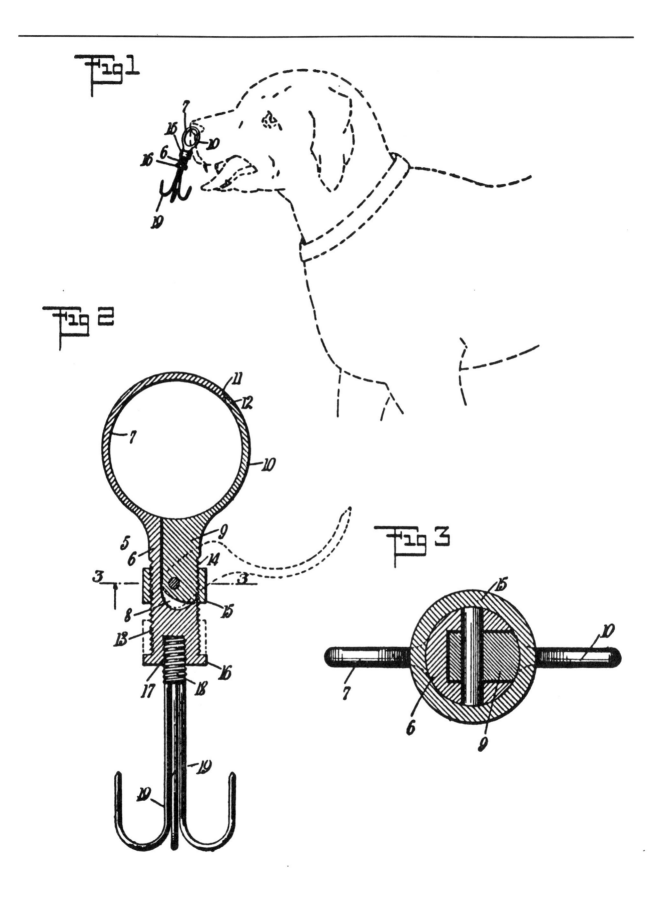

Hen Disinfector
Patent No. 1,303,851 (1919)
George Carlson of New York, New York

This invention is a novel poultry disinfector, that is to say, an apparatus adapted to dust an insecticide powder, or to spray any other disinfectant, upon live poultry. The principal object of the present invention is to afford such an apparatus which will be wholly self-acting and requiring very little attention from the poultry keeper, and which as well will be simple in structure, durable, effective in action and convenient of use.

The illustrated embodiment of the present invention [Figs. 1–4] discloses a compact and portable apparatus which can be placed where desired, for example, in the entrance to the hen-house or a runway, and as will be seen it is adapted to operate in both directions, that is, on poultry coming and going so that they are sprayed at both front and rear.

When the hen jumps on the front part 12 of the platform, the rear or interior end causes the piston to rise thus forcing air from the cylinder through the pipe 24, so as to carry the powder from the hopper 21 to the nozzle 16, whence it is sprayed upon the fowl. When the air is under pressure part of it will pass through the pinhole 23 into the hopper thus stirring up the disinfectant, and such air forced into the hopper promptly returns thus insuring passage of a suitable amount of powder into the pipe 24 at each operation.

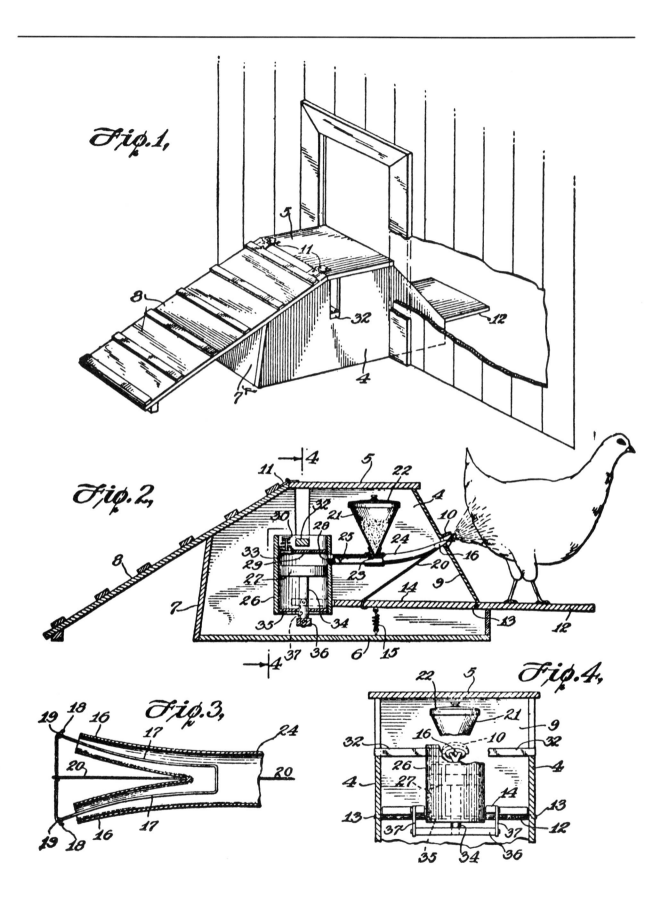

SEVEN

What Hath God Wrought!

Well before Samuel F. B. Morse flashed his famous first words—"What hath God wrought!"—over the electric telegraph from Washington to Baltimore in May 1844, men had begun to understand the powers of harnessed electricity. In 1600 William Gilbert, an Elizabethan court physician, published his treatise *On the Magnet.* In 1752 Benjamin Franklin developed the theory of positive and negative electricity and performed his famous kite-and-key experiment. Luigi Galvani's anatomic studies led to the invention of the electric battery (in 1801) by Allesandro Volta, a fellow Italian. Michael Faraday of London devised an electric motor in 1821; ten years later, he and Joseph Henry of Albany, New York, discovered the principle of the dynamo, or electric generator, which was still being perfected at the end of the century. As early as 1835 there was an electric automobile—built by Thomas Davenport, a blacksmith from Brandon, Vermont.

But it was the telegraph that touched off the communications explosion that is still being felt today. An impoverished portrait painter and photographer, Morse conceived the idea of transmitting intelligence by way of an electrically charged wire. He developed the "telegraphic alphabet"—the precursor of the Morse code—and in 1837 conducted his first successful test. For five years he petitioned Congress to appropriate funds for a long-distance telegraph line, and for five years he was rebuffed; one member of the House proposed an amendment to set aside half the requested funds for "experiments in mesmerism." Morse persisted, and was finally awarded a grant of $30,000 to string a wire across the forty miles between Baltimore and Washington. Unlike some other great inventors, Morse lived to see his telegraph span the continent and make the world itself a much smaller place (p. 140).

Alexander Graham Bell had similar good fortune. Born in Edinburgh in 1847 and trained as a doctor, Bell became professor of vocal physiology at Boston University and was quickly absorbed by the possibilities of transmitting musical notes by electricity. In 1876, almost by accident, he summoned his assistant, then several rooms away, with the first message transmitted by telephone: "Mr. Watson, come here—I

> Genius is mainly an affair of energy.
> Matthew Arnold,
> Essays in Criticism

want you!" (p. 142). The "Bell system" followed soon after, and before the turn of the century one in every fifty Americans had a unit. By age thirty Bell was both rich and famous.

Perhaps no American inventor is more well known than Thomas Edison. Born in Milan, Ohio, in 1847, he received his first patent in 1868 for an automatic vote recorder. Within a short time thereafter he invented a stock ticker, a duplex telegraph, and a carbon transmitter. By 1876 Edison had set up his famous laboratories in Menlo Park, New Jersey, where in 1877 he invented the phonograph (p. 144), and in 1879 the incandescent lamp (p. 146)—the first major advance in lighting since the gas lamp (1801), and the last until the fluorescent lamp (1934). Edison also devised the first central power station (1882) and motion-picture camera (1889).

The wireless radio was introduced in 1896 by Guglielmo Marconi (half-Italian, half-British) at age twenty-one. Five years later radio messages were being sent across the Atlantic. Other revolutionary devices for the transmission of sound included the magnetic tape recorder (1899), sonar (1916), the portable radio (1921, p. 154), the transistor (1948), and long-playing records (1948).

Electronic transmission of visual images was the development of many inventors, most of whom were applying earlier discoveries. The first television pictures appeared in 1926, and by 1939 regular programming had begun. A color-television patent was issued in 1935 (p. 156). The first videotape recorder was patented in 1956. Radar was invented in 1935, the computer in 1946, the laser beam in 1960, and the first communications satellite was sent into orbit in 1962.

Electricity has always had a certain mystique about it—which perhaps explains why John Bunyan Campbell was awarded a patent for his electric poison extractor (p. 152)—but its practical applications have proven so open-ended that it has been virtually impossible for the world to keep up with the technology, or for futurists to envision useful things beyond the realm of distinct possibility.

Telegraph Code
Patent No. 1,647 (1840)
Samuel F. B. Morse of New York, New York

Be it known that I have invented a new and useful machine and system of signs for transmitting intelligence between distant points by the means of a new application and effect of electro-magnetism in producing sounds and signs, or either, and also for recording permanently by the same means, and application, and effect of electro-magnetism, any signs thus produced and representing intelligence, transmitted as before named between distant points; and I denominate said invention the "American Electro-Magnetic Telegraph."

It consists of the following parts—first, of a circuit of electric or galvanic conductors from any generator of electricity or galvanism and of electro-magnets at any one or some points in said circuit [see Figs. 1–5]; second, a system of signs by which numerals, and words represented by numerals, and thereby sentences of words, as well as of numerals . . . are communicated to any one or more points in the before-described circuit; third, a set of type adapted to regulate the communication of the above mentioned signs, also cases for convenient keeping of the type and rules in which to set and use the type; fourth, an apparatus called the "straight port-rule," and another called the "circular port-rule," each of which regulates the movement of the type when in use; fifth, a signal-lever which breaks and connects the circuit of conductors; sixth, a register which records permanently the signs communicated at any desired points in the circuit; seventh, a dictionary or vocabulary of words to which are prefixed numerals for the uses hereinafter described; eighth, modes of laying the circuit of conductors.

The sign of a distinct numeral, or of a compound numeral when used in a sentence of words or of numerals, consists of a distance or space of separation between the characters of greater extent than the distance used in separating the characters that compose any such distinct or compound numeral.

Signs of letters consist in variations of the dots, marks, and dots and lines, and spaces of separation of the same formation as compose the signs of numerals, varied and combined differently to represent the letters of the alphabet.

The sign of a distinct letter, or of distinct words, when used in a sentence, is the same as that used in regard to numerals. . . .

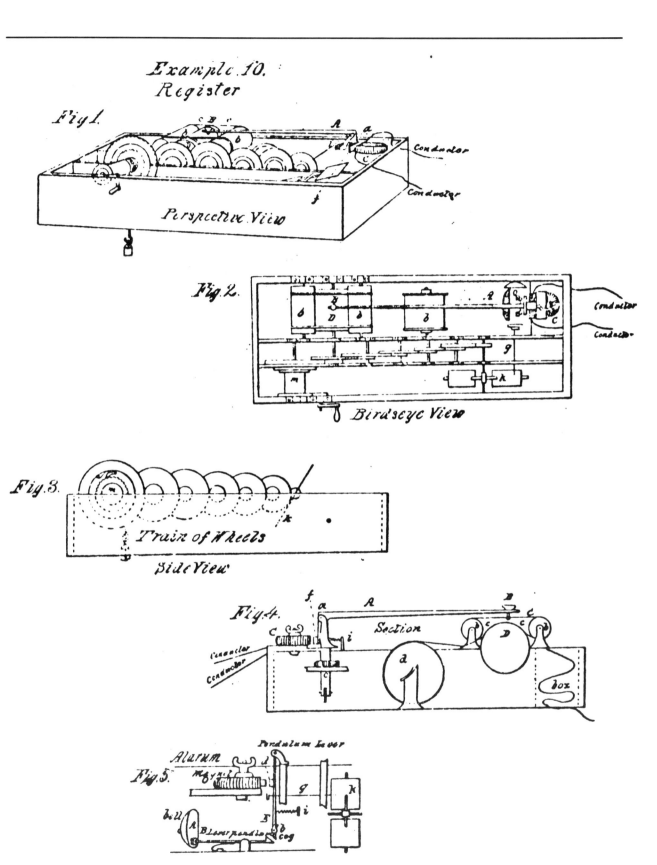

Telephone
Patent No. 174,465 (1876)
Alexander Graham Bell of Salem, Massachusetts

My present invention consists in the employment of a vibratory or undulatory current of electricity [see Fig. 4] in contradistinction to a merely intermittent or pulsatory current, and of a method of, and apparatus for, producing electrical undulations upon the line-wire.

The advantages I claim to derive from the use of an undulatory current in place of a merely intermittent one are, first, that a very much larger number of signals can be transmitted simultaneously on the same circuit; second, that a closed circuit and single main battery may be used; third, that communication in both directions is established without the necessity of special induction-coils; fourth, that cable dispatches may be transmitted more rapidly than by means of an intermittent current or by the methods at present in use; for, as it is unnecessary to discharge the cable before a new signal can be made, the lagging of cable-signals is prevented; fifth, and that as the circuit is never broken a spark-arrester becomes unnecessary.

When two or more instruments of different pitch are simultaneously caused to vibrate, all the instruments of corresponding pitches upon the circuit are set in vibration, each responding to that one only of the transmitting instruments with which it is in unison. Thus the signals of A, Fig. 6, are repeated by A^1 and A^2, but by no other instrument upon the circuit; the signals of B^2 by B and B^1; and the signals of C^1 by C and C^2—whether A, B^2, and C^1 are successively or simultaneously caused to vibrate. Hence by these instruments two or more telegraphic signals or messages may be sent simultaneously over the same circuit without interfering with one another.

I desire here to remark that there are many other uses to which these instruments may be put, such as the simultaneous transmission of musical notes, differing in loudness as well as in pitch, and the telegraphic transmission of noises or sounds of any kind.

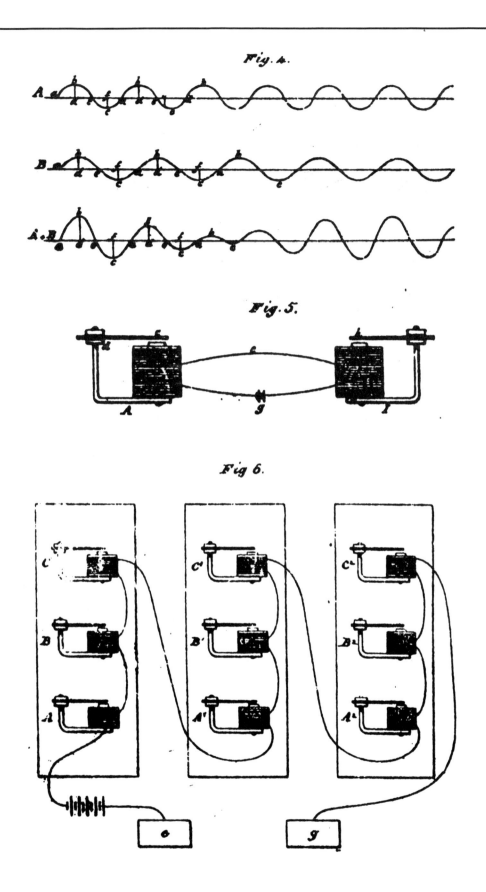

Phonograph
Patent No. 200,521 (1878)
Thomas Alva Edison of Menlo Park, New Jersey

I have discovered, after a long series of experiments, that a diaphragm or other body capable of being set in motion by the human voice does not give, except in rare instances, superimposed vibrations, as has heretofore been supposed, but that each vibration is separate and distinct, and therefore it becomes possible to record and reproduce the sounds of the human voice.

The object of this invention is to record in permanent characters the human voice and other sounds.

The invention consists in arranging a plate, diaphragm, or other flexible body capable of being vibrated by the human voice or other sounds, in conjunction with a material capable of registering the movements of such vibrating body by embossing or indenting or altering such material, in such a manner that such register-marks will be sufficient to cause a second vibrating plate or body to be set in motion, and thus reproduce the motions of the first vibrating body.

In the drawings, Figure 1 is a vertical section and Fig. 2 is a plane.

A is a cylinder having a helical indenting-groove cut from end to end—say ten grooves to the inch. Upon this is placed the material to be indented, preferably metallic foil. This drum or cylinder is secured to a shaft, X.

L is a tube, provided with a longitudinal slot, and it is rotated by the clock-work at M, or other source of power.

The shaft X passes into the tube L, and it is rotated by a pin, 2, secured to the shaft, and passing through the slot on the tube L, the object of the long slot being to allow the shaft X to pass endwise through the center or support P by the action of the screw on X. At the same time that the cylinder is rotated it passes toward the support O.

B is the speaking-tube or mouth-piece, which may be of any desired character, so long as proper slots or holes are provided to re-enforce the hissing consonants.

Upon the end of the tube or mouth-piece is a diaphragm, having an indenting-point of hard material secured to its center, and so arranged in relation to the cylinder A that the point will be exactly opposite the groove in the cylinder at any position the cylinder may occupy in its forward rotary movement.

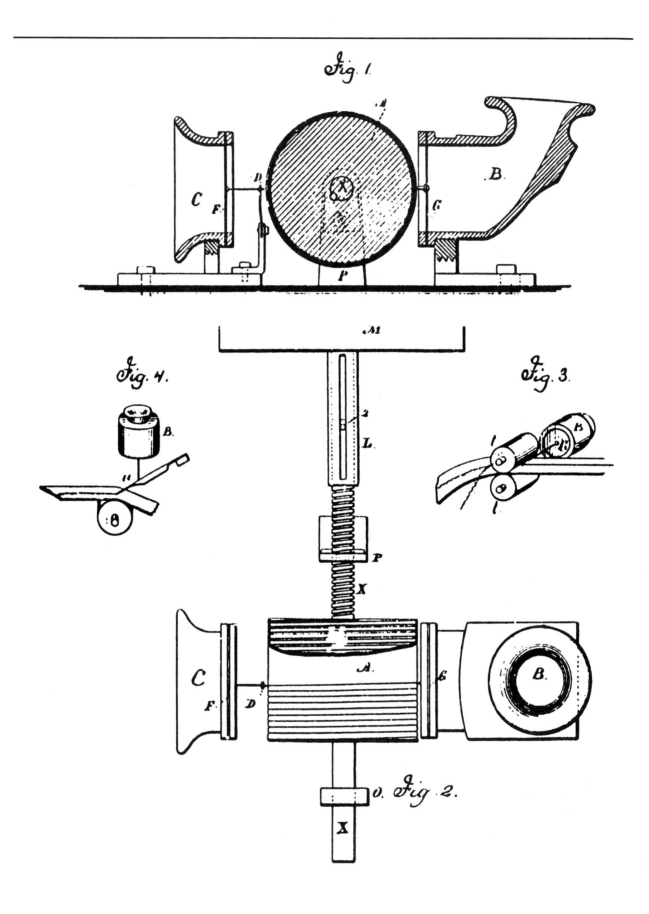

Light Bulb
Patent No. 223,898 (1880)
Thomas Alva Edison of Menlo Park, New Jersey

Heretofore light by incandescence has been obtained from rods of carbon of one to four ohms resistance, placed in closed vessels, in which the atmospheric air has been replaced by gases that do not combine chemically with the carbon. The vessel holding the burner has been composed of glass cemented to a metallic base. The connection between the leading-wires and the carbon has been obtained by clamping the carbon to the metal. The leading-wires have always been large, so that their resistance shall be many more times less than the burner, and, in general, the attempts of previous persons have been to reduce the resistance of the carbon rod. The disadvantages of following this practice are that a lamp having but one to four ohms resistance cannot be worked in great numbers in multiple arc without the employment of main conductors of enormous dimensions; that, owing to the low resistance of the lamp, the leading-wires must be of large dimensions and good conductors, and a glass globe cannot be kept tight at the place where the wires pass in and are cemented; hence the carbon is consumed, because there must be almost a perfect vacuum to render the carbon stable, especially when such carbon is small in mass and high in electrical resistance.

The object of this invention is to produce electric lamps giving light by incandescence, which lamps shall have high resistance, so as to allow of the practical subdivision of the electric light.

The invention [Figs. 1–3] consists in a light-giving body of carbon wire or sheets coiled or arranged in such a manner as to offer great resistance to the passage of the electric current, and at the same time present but a slight surface from which radiation can take place.

The invention further consists in placing such burner of great resistance in a nearly perfect vacuum, to prevent oxidation and injury to the conductor by the atmosphere. The current is conducted into the vacuum-bulb through platina wires sealed into the glass.

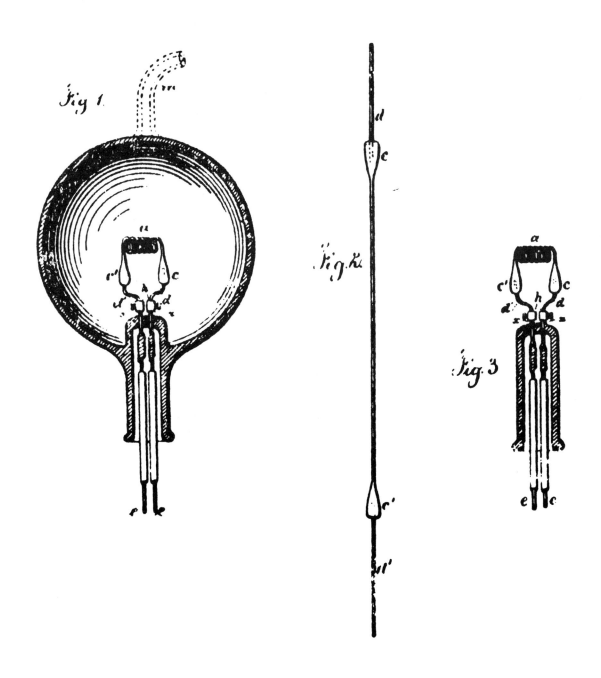

Electric Burglar Alarm
Patent No. 240,939 (1881)
Samuel S. Applegate of Camden, New Jersey

In Fig. 1 A represents an ordinary battery, of which x is the positive and g the negative wire, the latter extending directly from the battery to one post of a magneto-electric alarm, B. The positive wire x, however, is discontinued for an interval between the battery and the alarm for the introduction into the circuit of a mat, D. This mat is constructed of narrow strips of wood or other suitable material backed with canvas or like flexible fabric, and on the under side of the mat are certain spring-plates, which complete the circuit and cause the sounding of the alarm when pressure is exerted upon any portion of the mat—such, for instance, as would be caused by a person stepping upon it—the circuit being broken and the sounding of the alarm stopped when the pressure is removed. This mat was therefore adapted for use only on an open circuit.

One of the principal advantages of using the mat on a closed circuit is that it may be connected with the wires of the district-telegraph lines now in operation in most of the large cities of the United States, these lines being worked on closed circuit.

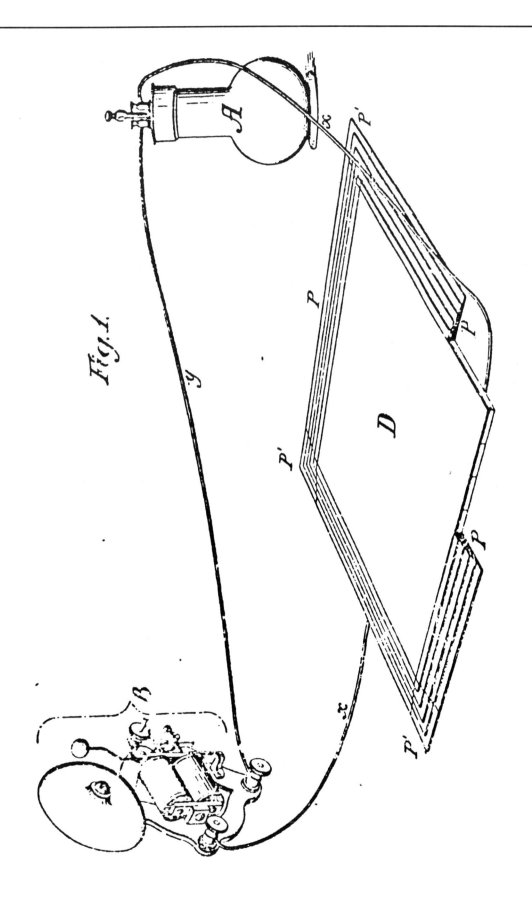

Fig. 1.

Electromagnetic Motor
Patent No. 381,968 (1888)
Nikola Tesla of New York, New York

The practical solution of the problem of the electrical conversion and transmission of mechanical energy involves certain requirements which the apparatus and systems heretofore employed have not been capable of fulfilling. Such a solution, primarily, demands a uniformity of speed in the motor irrespective of its load within its normal working limits. On the other hand, it is necessary, to attain a greater economy of conversion than has heretofore existed, to construct cheaper and more reliable and simple apparatus, and, lastly, the apparatus must be capable of easy management, and such that all danger from the use of currents of high tension, which are necessary to an economical transmission, may be avoided.

I secure, first, a uniform speed under all loads within the normal working limits of the motor without the use of any auxiliary regulator; second, synchronism between the motor and generator; third, greater efficiency by the more direct application of the current, no commutating devices being required on either the motor or generator; fourth, cheapness and economy in maintenance; fifth, the capability of being very easily managed or controlled; and, sixth, diminution of danger from injury to persons and apparatus. [See Figs. 13, 15, and 16.]

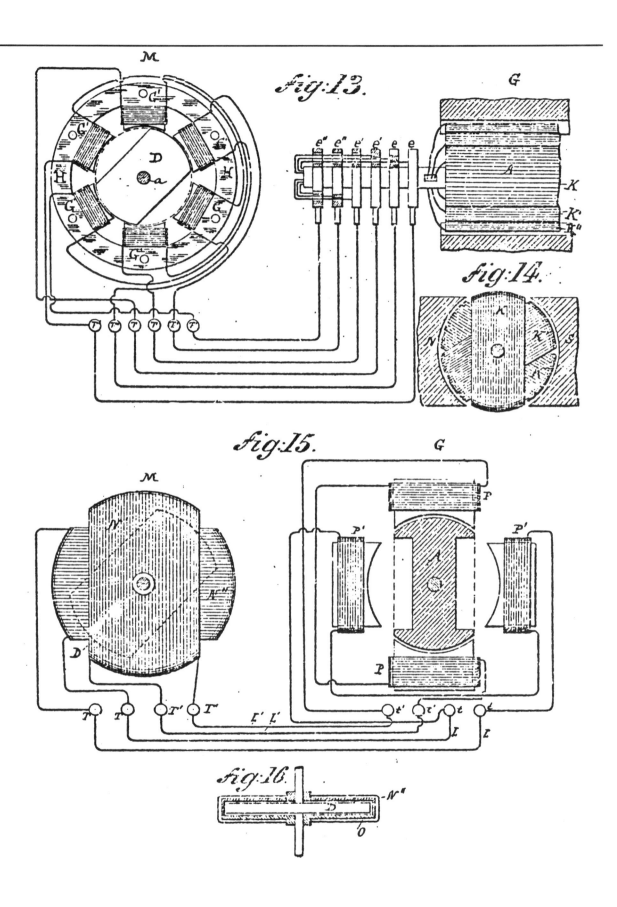

Electric Poison Extractor
Patent No. 606,887 (1898)
John Bunyan Campbell of Cincinnati, Ohio

Fig. 1 is a view in perspective of a male subject or patient seated in a chair, the electric battery, and the conducting-wires leading from the electric battery to the positive and negative plates applied to the back of the neck of the patient and at the same time to the bare feet of the patient or person receiving treatment.

For vegetable poisons I employ a vegetable receiver instead of a mineral or copper one, and for animal poisons I use an animal receiver, such as raw meat, the device being capable of use with the mineral, vegetable, or animal receivers without further change then to equip it with the kind of receiver applicable to the kind of poison desired to be extracted or removed from the human system.

The application of the different receivers is made to the negative electrode, and the positive electrode is applied to any suitable part of the body. When the current is turned on, it will run down from the neck or other suitable place through the patient's body and will pull or draw out the poison at the negative pole and deposit it on the copper plate. From six to eight treatments of a half an hour each in duration will generally extract all the poison of whatever kind it may be, and the copper plate will show as bright and clear as it was at first.

Portable Radio
Patent No. 1,386,840 (1921)
Reginald Charles Clinker of Rugby, England

It is known that by the use of sensitive detectors and rectifiers now available radio signals can be received without any aerial wire by means of a coil or coils set up in a suitable position relatively to the direction of the received waves.

The object of my invention is to render such receiving apparatus more readily portable and more compact. I have found that it is possible to make the frame upon which the receiving coil is wound contain within it the instruments (such as condensers, resistances, batteries and detectors) without any very serious damping of the received oscillations due to the presence of metal bodies within the coil. The coil can be mounted within a wooden casing provided with a handle and this arrangement results in a very compact and portable apparatus which occupies a small amount of space.

Figure 1 is an elevation of the casing with the cover removed showing the coil and other apparatus; Fig. 2 is a plane view.

To receive signals condensor 20 is adjusted until the natural frequency of coil 4 and condenser 20 corresponds to that of the received waves. The coupling between the coils 4 and 14 is then adjusted by the turning of the latter upon its hinges. The line of direction along which the signals travel is found by turning the casing on its pivot 15 until the signals have no effect upon the instrument, in which position the pointer 18 would indicate the direction desired.

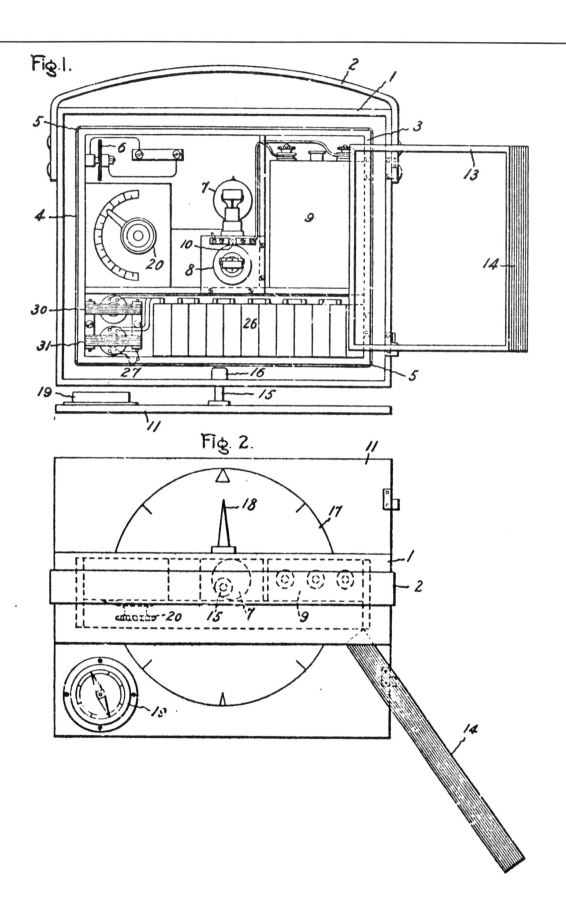

Fig. 1.

Fig. 2.

Color Television
Patent No. 2,013,162 (1935)
Harold J. McCreary of Chicago, Illinois

Images of objects have heretofore been transmitted through distances and the present invention relates primarily to this art. In the prior methods or systems, so far as I know, the object [whose] image was to be transmitted was scanned by means of mirrors, shutters, or other movable mechanical devices. The primary object of the present invention is to produce a system of transmission of images which does not depend upon movable mechanical elements of any kind excepting, of course, the movable members of generators which supply the necessary alternating currents.

In carrying out my invention [Fig. 1], I scan the image of the sending station with a cathode ray and simultaneously scan a sensitive or responsive surface at the receiving station to reproduce the image. In one of its aspects, therefore, my invention may be regarded as having for its object to produce synchronous deflection of cathode rays at the sending and the receiving stations.

So far as I know, no prior system has transmitted images in the natural colors of the objects whose images are to be transmitted. Viewed in one of its aspects, my invention may be said to have for one of its objects to produce a simple and novel system of transmitting images in natural colors. . . .

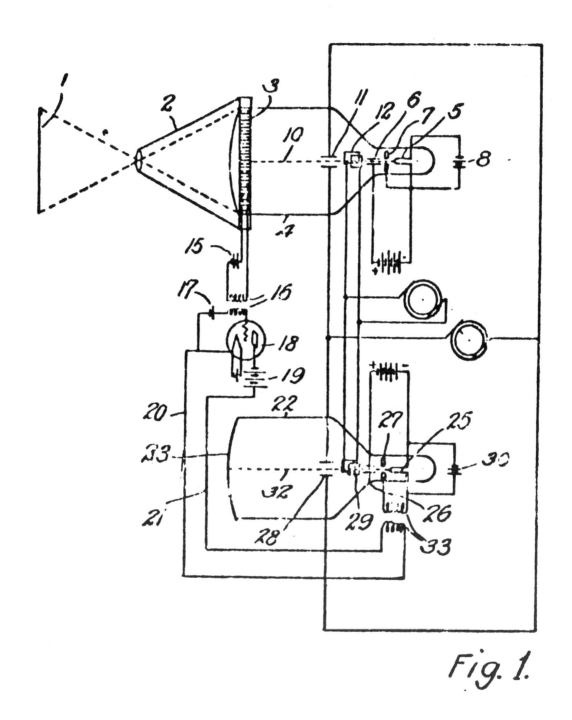

Fig. 1.

Illuminated Brooch
Patent No. 2,374,375 (1945)
James H. O'Donnell of Woburn, Massachusetts

My invention relates to illuminated brooches to be worn upon the clothing of a person for purposes of display, amusement and the like, and it has for its object to provide an inexpensive and efficient self-contained device of the character described which can be detachably fastened in position upon the clothing of the wearer and which shall be provided with illuminating means that can be lightened and extinguished at the will of the wearer.

The design of the device shown in Figs. 1, 2 and 3 is appropriate for use by young girls and other young people. For example, when worn by a young girl or woman who is approached by a young man seeking her acquaintance, she can signal encouragement to him by illuminating the green transparency to the exclusion of the red by adjustment of the lamp thereof, or should she desire to repel his attentions then she can illuminate the red transparency to the exclusion of the green by adjustment of the lamp of the same.

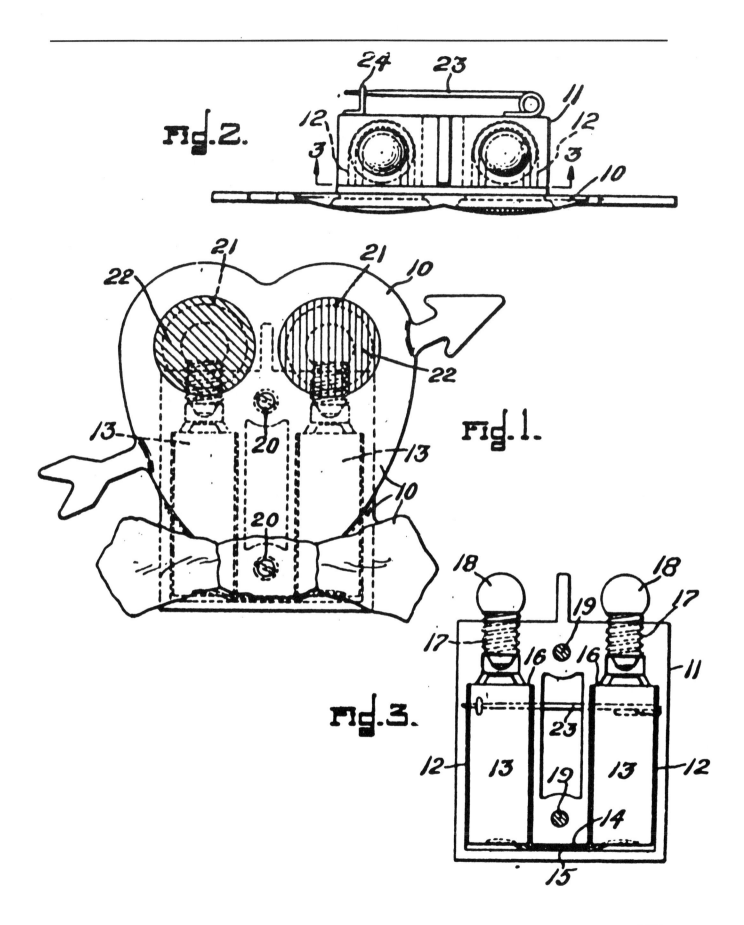

Calculator
Patent No. 2,668,661 (1954)
George R. Stibitz of Boonton, New Jersey

This invention relates to calculating methods and apparatus and particularly to systems of electrical control whereby the various mathematical operations required are performed with rapidity and without the use of counting wheels, gear trains and similar mechanical elements.

An object of the invention is to provide a rapidly operating and economical arrangement in which all calculating movements are confined to simple operations of electromagnetic circuit makers and breakers [Fig. 1].

Another object of the invention is the use of calculating methods which may be translated into simple and small movements of electromagnetic apparatus for the control of electrical circuits.

Another object is to achieve simplicity and economy through the avoidance of the use of complicated mechanical movements and expensive precision built machine elements.

A further object is to provide a calculating mechanism operated by remote control and particularly a common calculating device which may be operated from any one of a plurality of distant control stations.

Another object of the invention is to provide electrical means for performing the simple algebraic operations of addition, subtraction, division and multiplication.

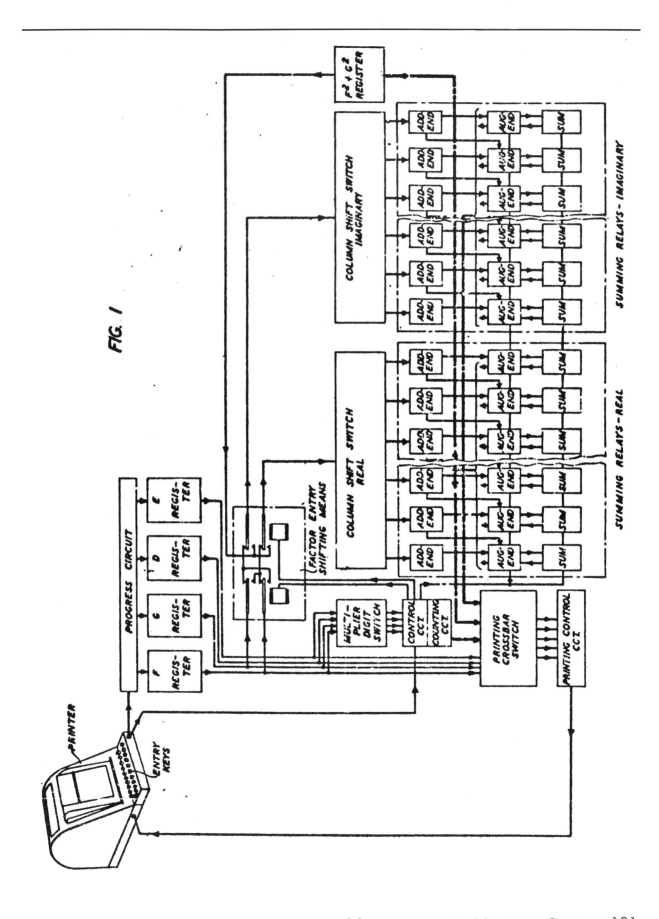

EIGHT

Loco Motion

Why is it, the question has been asked, that people have always wanted to ride faster, sail farther, soar higher? Part of the answer could be simple social and commercial necessity, but another reason would undoubtedly be the sheer joy that comes from tinkering with machines that move.

Take Bob Shapleigh, for example, of Dover-Foxcroft, Maine. Mr. Shapleigh has built the world's first and only car that runs—really runs—on chicken fat. He has not even sought a patent for his proven gas-mocker, because for him the end-all and be-all is that it works. Let somebody else perfect it and make a profit.

Much of the early railway technology was developed by British tinkerers who in 1830 brought together two important inventions for the first time: the steam locomotive and the railroad track. "Iron horses" had finally enabled men to travel faster than animals. Urban tramways followed soon thereafter (their cable drums driven by steam engines), as did the first metropolitan subway (in London, in 1863). The first electric trolley did not appear until 1881—but it preceded the modern bicycle (chain-driven, with two-sized wheels and pneumatic tires) by almost a decade.

Meanwhile various European inventors were actively working on carburetors and internal-combustion engines, both of which had prototypes dating back to the 1860s. By 1887 Karl Benz of Mannheim, Germany, was making and selling three-wheeled, gasoline-powered horseless carriages, capable of achieving an average speed of eight and a half miles per hour. At the same time Gottlieb Daimler, of Württemberg, Germany, was building the first motorcycle. Neither man knew what the other was doing—nor did they ever meet—but until 1920 the motorcycle seriously rivaled the motorcar in popularity.

By the time Henry Ford patented his carburetor (in 1898; see p. 172) and designed the first large-scale assembly line for mass production of his Model T "Tin Lizzies" (1908), the pneumatic tire (1888), diesel engine (1893), drum brake (1902), four-wheel suspension (1905), and safety glass (1905) had already been invented. The first electric headlamps came in 1910, traffic lights in 1914, windshield wipers in 1920, parking meters (p. 180) in 1938, and automatic transmissions in

> The advancement of the arts, from year to year, taxes our
> credulity and seems to presage the arrival of that period
> when human improvement must end.
> *Henry Ellsworth*
> (first Commissioner of Patents)

1939. Everything since has been largely gadgetry and cosmetics.

In the early nineteenth century, the primary means of commercial transportation was by riverway. Both John Fitch of Philadelphia and James Rumsey of Bath, Virginia, built working steamboats in the 1780s that sailed on the Delaware and Potomac rivers, but never made commercial successes. "I know of nothing so perplexing and vexatious to a man of feelings," wrote Fitch, "as a turbulent Wife and Steamboat building." It was thus left for Robert Fulton of Lancaster, Pennsylvania, to achieve fame and fortune from steam power. But even his *Clermont* (1807)—before it sailed 150 miles of the Hudson in a record thirty-two hours—was dubbed "Fulton's Folly."

The next great advance in sea travel was not H. R. Rowland's water walkers (p. 188) but the submarine, invented by an Irish schoolmaster named John Holland in 1875. Enrico Forlanini, an Italian, built a small hydrofoil boat in 1906. The first hovercraft came in 1955, the same year as a nuclear-powered submarine, the USS *Nautilus*.

Manned flight began with a balloon ride over Paris in 1783, but it was more than a century before the first airship—one capable of being propelled and steered horizontally—was built. In 1884 *La France*, piloted by Charles Renard and Arthur Krebs, made a round trip of about five miles at the grand speed of fourteen and a half miles per hour. Graf Zeppelin's dirigible (1899) quickly became very popular (p. 194). Various nineteenth-century attempts at winged flight, however, were unsuccessful (see p. 192 for example).

Wilbur and Orville Wright of Dayton, Ohio, were bicycle makers both fascinated with wings. They built several gliders, which they successfully flew during 1900–1903 on the windy shores of Kitty Hawk, North Carolina. In 1903 they added an engine and propellers, and on December 17 made four brief winged flights. It was the first time man had flown in a heavier-than-air machine.

The subsequent rapid progress of aviation has been little short of fantastic, from the development of the helicopter (1907) to the warplane (1915), the rocket (1926) to the jet engine (1939), and the first transatlantic flight (1927) to extended travels in space (1957). For the many inventors involved, the sky was no limit.

Air Brakes
Patent No. 88,929 (1869)
George Westinghouse, Jr., of Schenectady, New York

My invention relates to the construction of a power car-brake for railway-cars or other like vehicles, to be operated by compressed air or other elastic compressible fluid; and the nature of it consists, first, in the use of an auxiliary engine for compressing the air in a reservoir, from which it is to be conducted by suitable pipes and applied to operating the brakes, and also for pumping feed-water into the boiler, either or both; second, in the construction of a reservoir for storing up the power to be derived from air or other elastic fluid under compression; third, in the construction and combination of devices by which the power thus communicated to the piston of the brake-cylinder may be from it applied to operating an ordinary hand-brake; fourth, in the construction of an improved coupling for uniting the brake-pipe of contiguous cars, so made that when coupled they shall be always open for the passage of air to the brake-cylinders, but if uncoupled, when the brakes are down, the pressure of the air in the pipes will close them.

Figure 1 shows by a side elevation, partly in section, my improvement as mounted on and applied to an ordinary platform railroad car. Fig. 2 is a sectional plan view as formed by a horizontal plane passing through the brake-cylinder, just below the body of the car. Fig. 3 is a vertical, and Fig. 4 is a cross section of the three-way cock by which I admit or cut off the supply of air to the brake. Fig. 5 is an outside view, and Fig. 6 is a longitudinal section of my improved coupling for uniting the brake-pipes of contiguous cars.

The particular advantages connected with the apparatus described, in addition to those above referred to, are that the brakes are under the control of the engineer, and can be instantaneously applied at any time, and with any degree of power within the strength of the machinery employed, and the brakes can be as instantaneously loosened. They are simple in construction, cheaply made, and can be applied to and used in connection with or used without the ordinary hand-brakes. In case of accident the brakes may be instantaneously applied, and kept "on" till the whole train, or each separate car, if the car-couplings break, be brought to a complete stand.

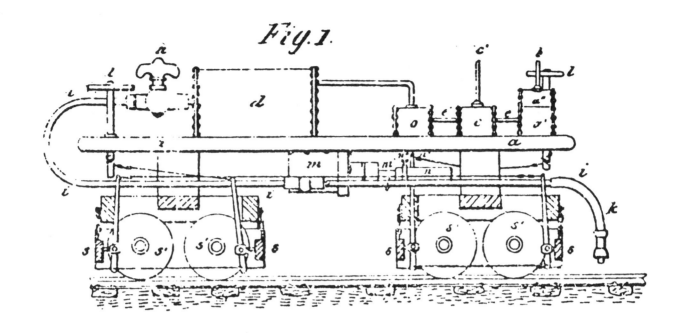
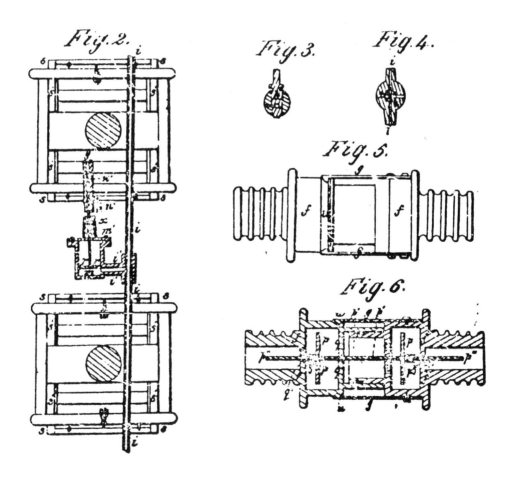

Steam Railroad Cattle Prod
Patent No. 292,504 (1884)
LaFayette W. Page of Shreveport, Louisiana

This invention relates to an attachment for locomotives, to be used for frightening horses and cattle off the track.

By means of the rod I [Figs. 1–3] the stop-cock F may be opened, thus permitting the water to escape from the boiler through the nozzle D, through which it is driven, by the steam-pressure in the boiler, with a great degree of force, and to a considerable distance, so that it may be employed for frightening horses and cattle off the track. By means of the rod L the nozzle may be adjusted so as to throw the stream of water in other than a straight line, so that the device may be advantageously used on curves.

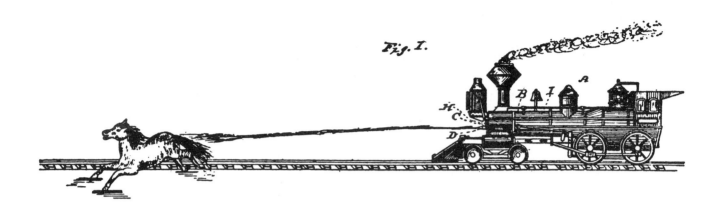
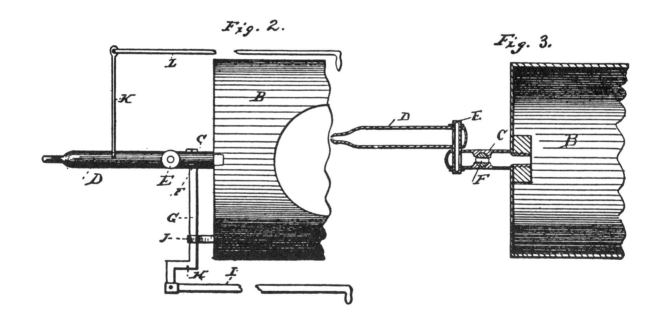

Mousetraps and Muffling Cups 167

Railroad Hammock
Patent No. 400,131 (1889)
Herbert Morley Small of Baldwinsville, Massachusetts

This invention relates to an improvement in hammocks, and has special reference to a hammock so constructed that it can be slung from and used with the backs of the seats of ordinary railway passenger-cars.

The invention has for its object to provide a means whereby passengers who are obliged to travel in ordinary passenger-cars at night may be able to sleep with ease and comfort.

Figure 1 illustrates the invention in use in an ordinary passenger-car. Fig. 2 is a view of a hammock constructed in accordance with this invention, and Fig. 3 represents the end of a hammock provided with a modified form of attachment and legs and feet support.

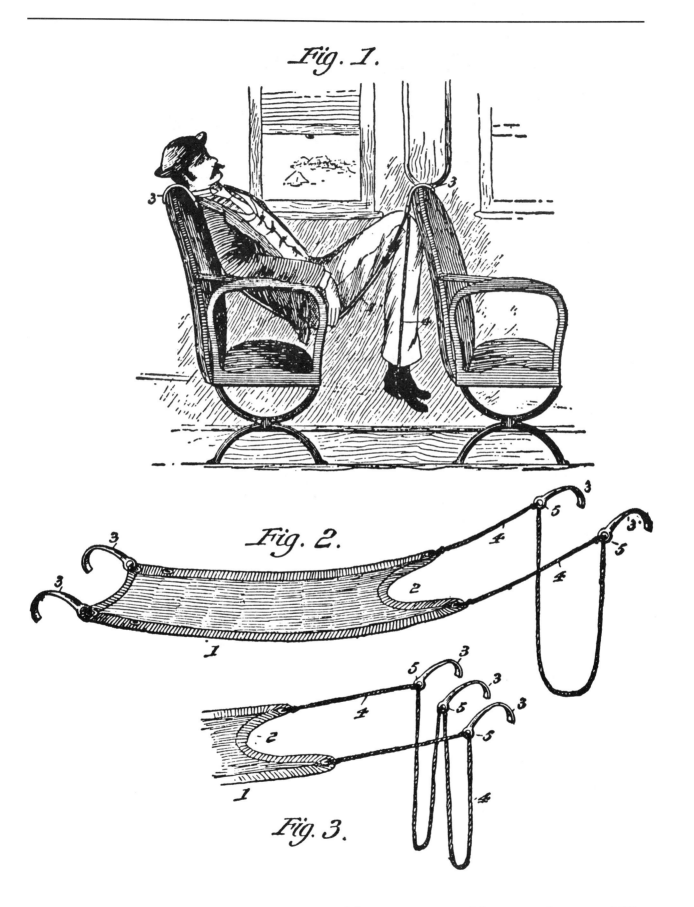

MOUSETRAPS AND MUFFLING CUPS

Crashproof Railroad Cars
Patent No. 536,360 (1895)
Henry Latimer Simmons of Wickes, Montana

When one train meets or overtakes another train, one train will run up the rails E, carried by the other train, and will run along the rails E and descend onto the rails A at the other end of the lower train, as shown in Fig. 1.

The trains have the inclined lower ends of their rails E adjusted at different distances from the rails A in a pre-arranged manner so that the ends of the rails E of successive trains are not exactly upon the same level when they meet. The train having the ends of its rails E higher above the rails A than those of the train it meets will rise and run up on the rails E of the other train.

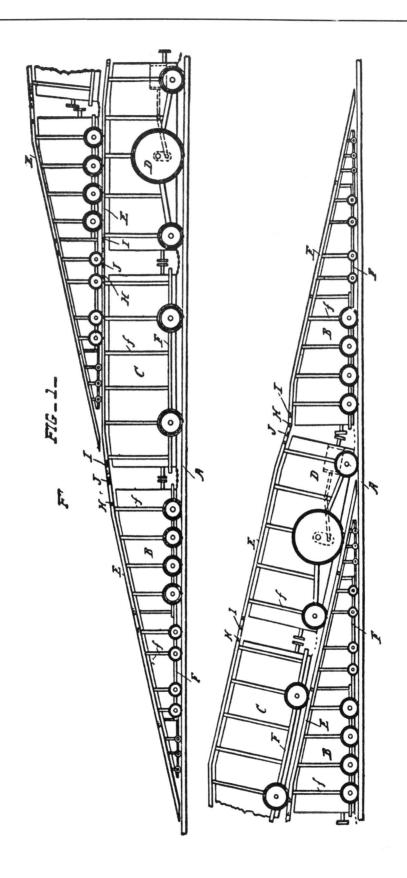

Carburetor
Patent No. 610,040 (1898)
Henry Ford of Detroit, Michigan

The invention relates to the construction of a carburetor especially designed for use in connection with gas or vapor engines; and it consists, first, in the construction of a device for feeding a fixed amount of the carbureting fluid; second, in the means for feeding that charge, with a requisite amount of air, into the explosion cylinder or chamber of the engine; and, third, in the construction, arrangement, and combination of the various parts. . . .

In the drawings, Figure 1 is a perspective view of my invention. Fig. 2 is a vertical central section therethrough. Fig. 3 is a horizontal section above the feed-disk. Fig. 4 is a diagram elevation showing the connections between the main tank containing the fluid-supply and the cup containing the cut-off as well as the air connections.

The operation of the device is as follows: The gasolene or other carbureting fluid will be fed from the tank into the cup and fill those cells or pockets in the feed-disk which are exposed or out from under the cut-off. Suitable mechanism on the engine actuates the shaft M, so as to bring the filled pockets into line with the pipes O and E at the time that the air is being drawn through the pipe L to the explosion-chamber, and the suction of this air will draw through the pipes E and C and with it the fluid in the cell, thus thoroughly spraying it to carburet the air, which carries it to the explosion-chamber in such a manner that when the spark is produced in that chamber the explosion will take place. By this arrangement I get a fixed charge of the carbureting fluid with an exceedingly simple mechanism for feeding it. As the cell thus relieved of its fluid is moved on beneath the cut-off G the air contained therein will rise through the fluid and find exit through the air-escape pipe back into the top of the tank, so that I may use a closed tank I, for just the proper amount of air will be supplied to take the place of the oil displaced.

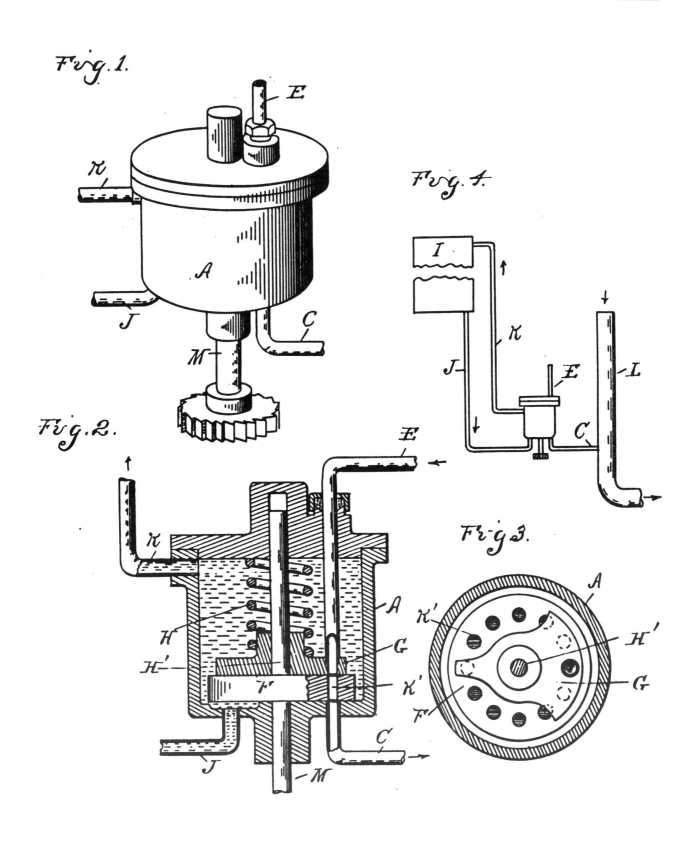

Automobile Chassis
Patent No. 686,046 (1901)
Henry Ford of Detroit, Michigan

The invention relates to a motor-carriage, and has for its object an improvement in the vehicle itself comprising an improved reach-rod and connections between the reach-rods and axles, an improved means of changing the angles between the front wheels and the carriage-body, and thereby changing the direction of the progressive motion of the vehicle.

In the drawings, Figure 1 is a bottom plan view of my improved vehicle, showing the rear axle, on which rests the frame of the engine, the gear-cases, and detached plan of the front axle and the levers used to change the vertical axis of the front wheels. Fig. 2 is a rear elevation of the rear axle with the motor thereon.

The wagon has two axles, and the forward axle instead of turning on a fifth-wheel is for most of its length fixed in a constant position, so as to be always parallel with the rear axle, except, however, that it is able to oscillate somewhat in a vertical plane parallel to that passing vertically through the rear axle. It can move with an oscillatory movement around its middle point. The amount of movement is not great but is sufficient to enable the carriage to accommodate itself to ordinary inequalities of the road.

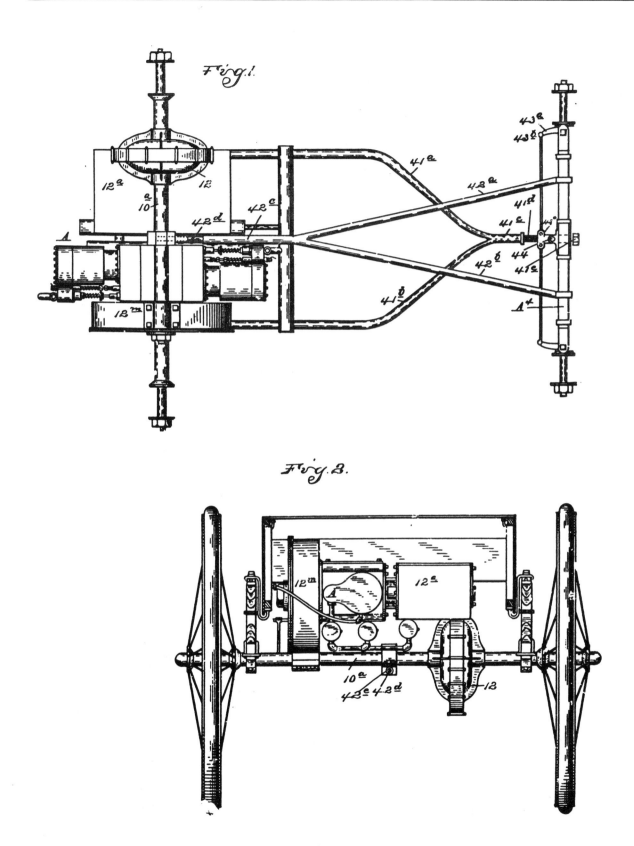

Horse Accessory for Cars
Patent No. 777,369 (1904)
Henry Hayes of Denver, Colorado

This invention relates to an attachment for motor-vehicles designed to prevent motor-vehicles from frightening the horses attached to horse-drawn vehicles upon the highways.

The object of the invention is to provide a figure of an animal, preferably a horse of approximately life-size, with means for attaching it to the front of a motor-vehicle in such manner that it may be propelled by the vehicle and present the appearance of a horse drawing the vehicle [Figs. 1 and 2].

In the head H of the figure I preferably form a chamber 9 to receive a lamp 10 of any preferred type, and above the chamber 9 a cover-plate 11 is hinged at 12. The cover-plate 11 comprises a perforated dome 13 for the escape of the products of combustion from the lamp 10, and near the lower margin the plate is provided with a lens 14, through which the rays from the lamp are directed upon the roadway. At either side of the head of the figure F there are preferably placed colored lenses 15, which correspond to the eyes of the animal.

The head H of the figure has hinged thereto a block 16, which represents the lower jaw of the animal, and a spring 17 is preferably provided by which the block is normally raised. Between the block 16 and the main portion of the head of the figure a bulb 18 is provided, and this bulb 18 has attached thereto a short flaring tube 19 or other device, through which air may be forced to produce a sound of sufficient intensity to serve as an alarm.

The action of the motor-vehicle attachment described is so plainly evident from a mere inspection of the drawings that a detailed description thereof appears to be unnecessary. As the figure is pivotally attached at the rear to the front axle of the motor-vehicle so as to have pivotal movement in a vertical plane and is provided in front with a swiveled supporting-wheel, the figure will pass readily over the ordinary highways without interfering with the travel of the vehicle, and the slight additional weight of the figure will not impede the progress of the vehicle to any considerable extent.

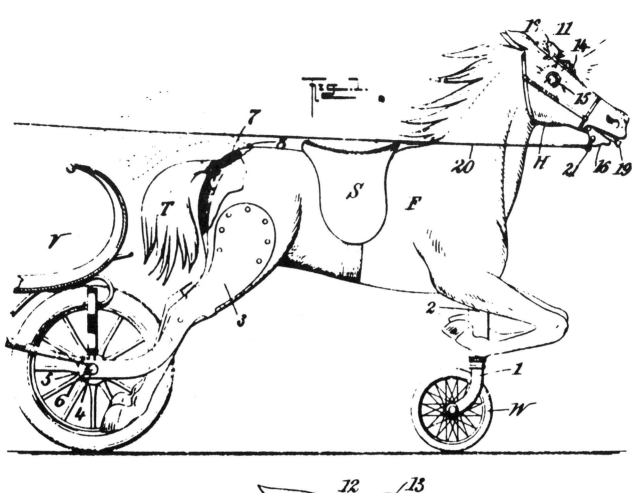
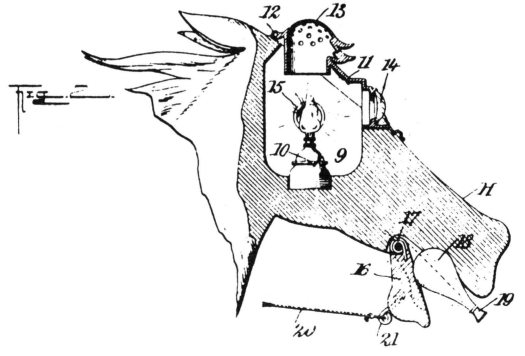

Improved Bicycle Seat
Patent No. 1,538,542 (1925)
Fred Blaje of Indianapolis, Indiana

It is a primary object of my invention to provide a bicycle or motorcycle saddle having a suitable cavity properly located to allow comfortable clearance for the private organs of the male rider, said saddle having also a channel adapted to allow clearance for the female rider's private organs, to prevent pressure at the opening of said organs due to the weight of the rider, and tending also to keep said organs in a naturally closed state, the sides of said channel being substantially parallel and bell-mouthed.

Referring to Fig. 1, a somewhat circular cavity 3 is formed on a medial line of the apparatus as a whole, at the junctions of said extension with said larger seating portion. Said cavity is of suitable size, shape and location to comfortably receive the private organs of a male rider and more particularly the testicle region of such rider. Said opening is of bell-mouthed formation, the bell-mouth character thereof being formed on the upper portion of said opening [Fig. 2]. Such bell-mouthed formation is particularly useful to the comfort of the male rider both during the riding act and also during the mounting or dismounting acts.

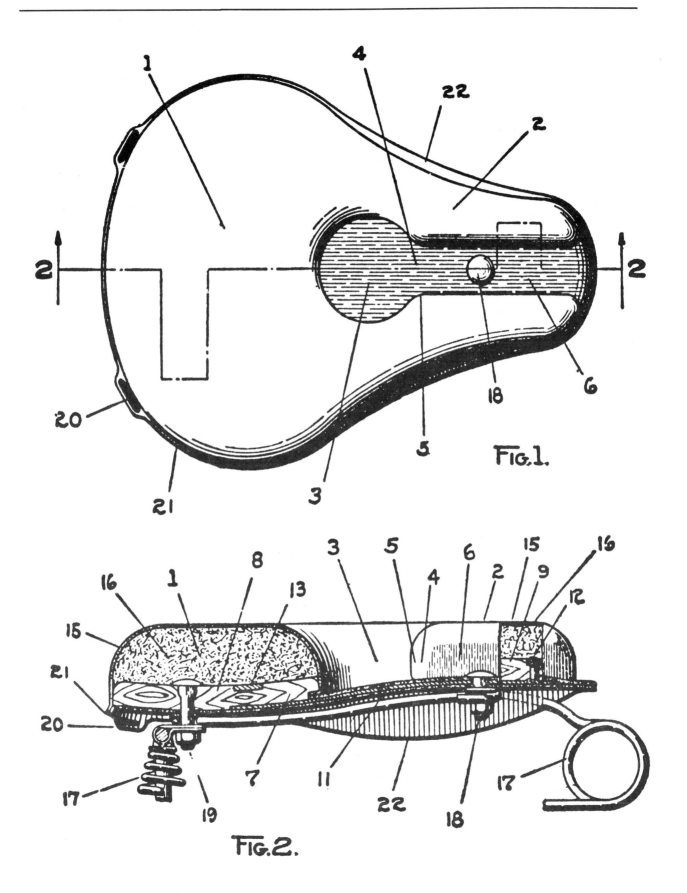

Parking Meter
Patent No. 2,115,024 (1938)
Charlie Klemt of San Antonio, Texas

The invention relates to coin or check-controlled meters used along curbs in cities, or at other locations, and of such construction as to be brought into operation by insertion of a coin or check and adapted to give some sort of an indication a predetermined time after insertion of the coin or check. When this indication is given, if the car parked by the meter has not been moved, another coin or check should be inserted for continuation of the parking privilege for a like period. If this be not done however, anyone looking at the meter is therefore of great advantage to traffic and police officers.

A further object is to provide a novel construction whereby if the person previously using the meter has not used his allotted time completely, the next user will neither gain nor lose thereby, for provision is made whereby any person inserting a coin and operating the control will effect resetting of the time-indicating means at zero position so that he will benefit for the entire period paid for by the coin or check but will not have any "left-over" time paid for by the previous user, either added to or subtracted from the time for which the last coin or check inserted has paid.

We will assume that one person has made use of the parking meter and has used say forty-five minutes of the allotted one hour. The next person, we will assume, enters the parking space as soon as the previous user leaves it. He deposits a coin or other check into the chute 18 [see Figs. 1 and 2] and pushes the plunger 54. The result is that the dog 33 which previously held the indicator 25 in the last position to which it was moved, is released, and said indicator being biased toward its zero position, automatically returns to said position upon release of said dog.

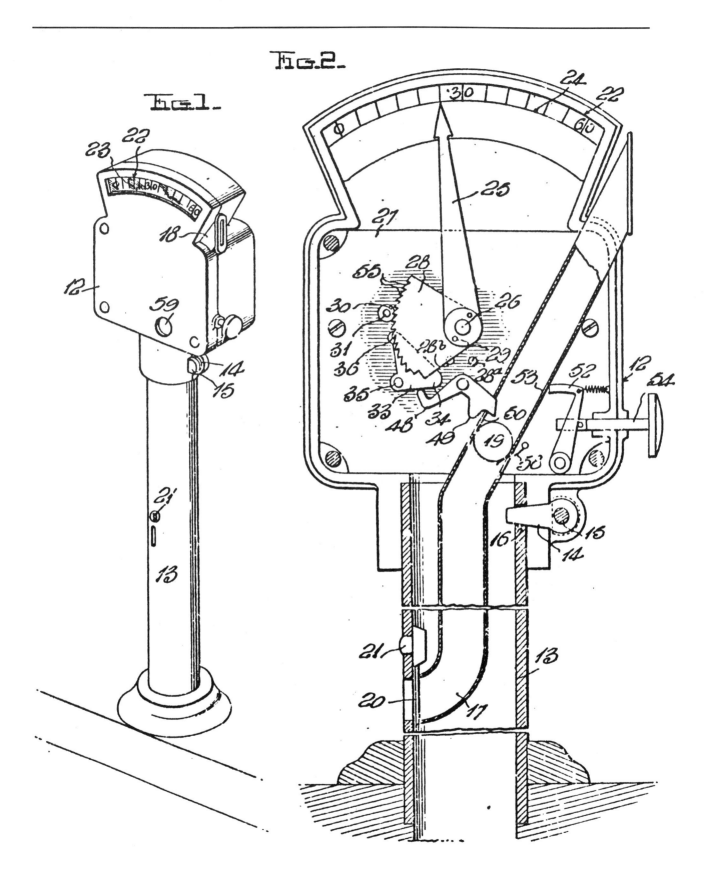

Horse Taillight
Patent No. 2,079,053 (1937)
John E. Torbert, Jr., of National City, California

One object of the invention is to provide a device of this character adapted to be supported and held in place by engagement with the tail of the animal and carrying safety reflectors disposed at opposite sides of the animal's tail so that when a person is riding a horse along a road at night and an automobile approaches the horse from the rear, the signals will be illuminated by reflecting light from the headlights of the automobile and thus permit the driver of the automobile to see that there is a horse ahead of him and eliminate danger of the automobile striking and injuring the horse.

Another object of the invention is to provide the attachment with a shield disposed at the back of the horse under the tail where it will serve as a wind deflector.

Another object of the invention is to so mount the shield that it may have movement towards and away from the horse as the horse rises or lowers its tail and to also provide a wiper mounted for oscillating movement across the shield and having an actuating member adapted to be connected with the horse's tail so that as the horse moves its tail to push off flies, the wiper will be moved back and forth across the windshield and the glass forming part of the shield kept clean.

This attachment, which may be referred to as a novelty or as a combined windshield and safety device for a horse or other animal, is to be applied to the horse, as shown in Figure 1, and when in place will attract the attention of persons attending a horse show or the like and afford a great deal of amusement. The device is also of practical value as it serves very effectively as a safety device to prevent accidents due to the driver of the automobile failing to see a horse ahead of him when driving during dark nights. While the device has been shown applied to a horse, it will be apparent that, by making it of the proper size, it may be applied to dogs and other animals. [See also Fig. 2.]

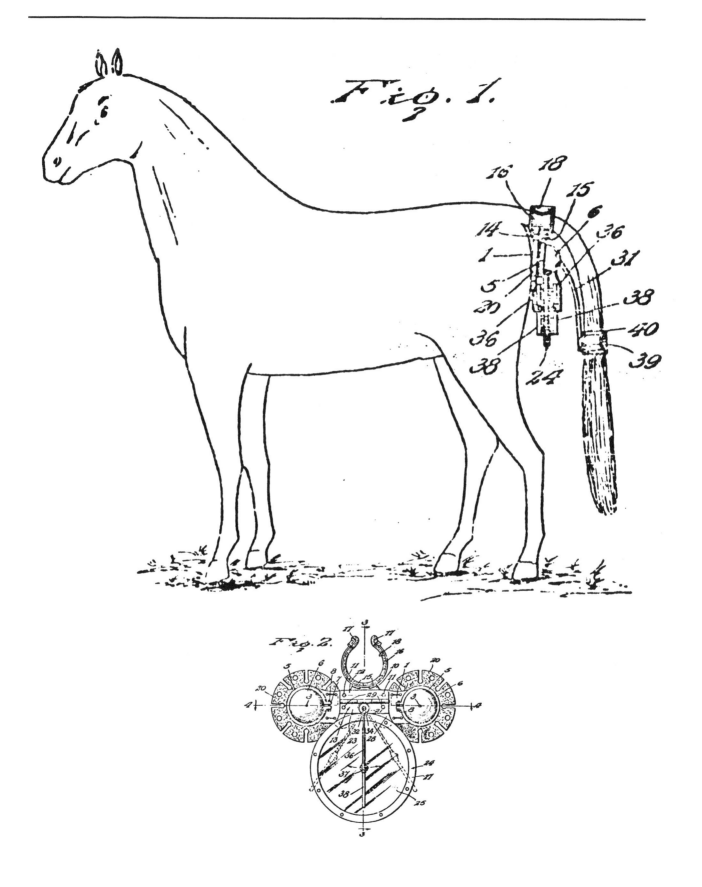

Adjustable Skateboard
Patent No. 3,235,282 (1966)
Louis D. Bostick of Granada Hills, California

A variety of skate board constructions have been in general use in coasting downgrade. Design changes made in recent years have increased the versatility of these devices and they have become attractive for use by adults as well as children. As heretofore made, such boards embody a rigid long main body and a pair of carriage units immovably secured to the underside of the board crosswise of its opposite ends.

By the present invention [Figs. 1–8] there is provided an improved and far more versatile skate board having an unusually wide range of operating characteristics and modes of use made available through the simple expedient of varying the length of the wheel base relative to the opposite ends of the board. By use of this expedient the wheel base may be made very short or very long and the wheel base proper may be shifted in either direction with respect to the midlength of the board. Each of the wide variety of wheel bases and positions imparts its own distinctive operating characteristics to the board and varies the skill required of the user. In some positions of the wheels the board is relatively easy to control while in others a very considerable amount of manipulative skill and dexterity is required. In some positions the board is relatively easy to steer along a complex path of travel whereas in others the board is more difficult to handle and requires greater dexterity and skill.

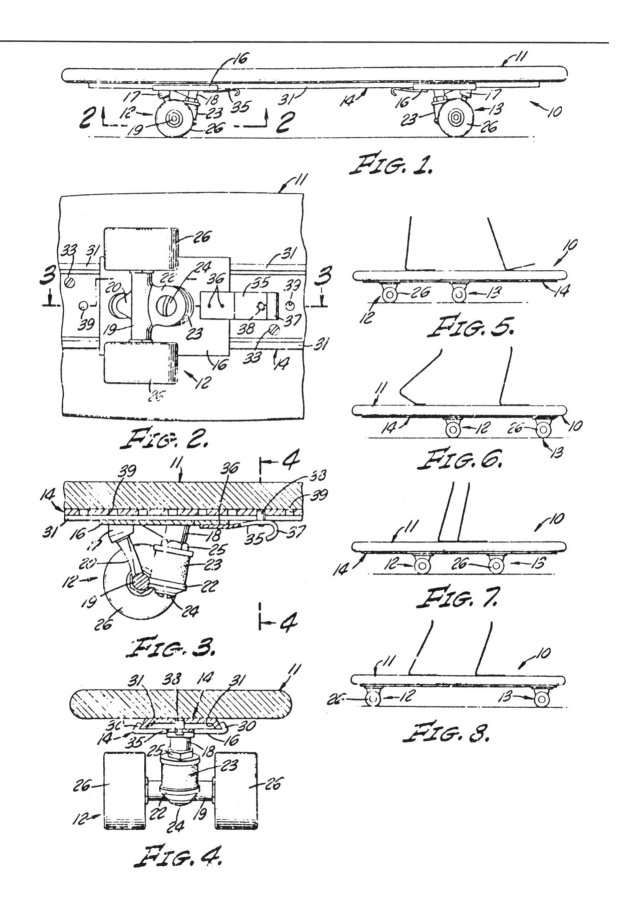

Buoyant Boat Chambers
Patent No. 6,469 (1849)
Abraham Lincoln of Springfield, Illinois

Be it known that I, Abraham Lincoln, of Springfield, in the County of Sangamon, in the State of Illinois, have invented a new and improved manner of combining adjustable buoyant air chambers with a steamboat or other vessel for the purpose of enabling their draught of water to be readily lessened to enable them to pass over bars, or through shallow water, without discharging their cargoes; and I do hereby declare the following to be a full, clear, and exact description thereof, reference being had to the accompanying drawings [Figs. 1–3] making a part of this specification. Similar letters indicate like parts in all the figures.

The buoyant chambers A, A, which I employ, are constructed in such a manner that they can be expanded so as to hold a large volume of air when required for use, and can be contracted, into a very small space and safely secured as soon as their services can be dispensed with.

Fig. 1 is a side elevation of a vessel with the buoyant chambers combined therewith, expanded;

Fig. 2 is a transverse section of the same with the buoyant chambers contracted.

Fig. 3 is a longitudinal vertical section through the centre of one of the buoyant chambers, and the box B, for receiving it when contracted, which is secured to the lower guard of the vessel.

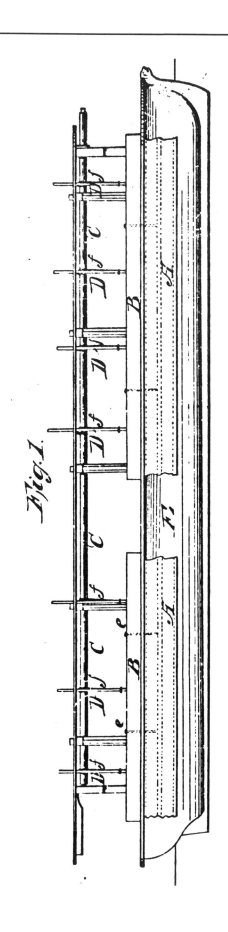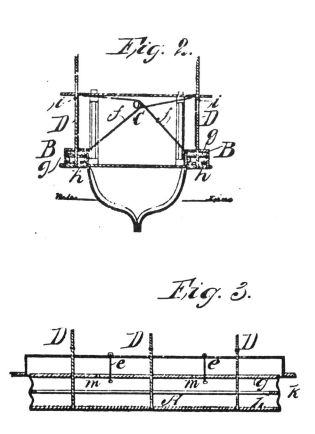

Water Walkers
Patent No. 22,457 (1858)
Henry R. Rowlands of Boston, Massachusetts

The operator standing in the steps c, c [Figs. 1 and 2], belays taut the forward floats lines and slackens the after ones; then, taking hold of the stanchions $H, H,$ he advances one foot forward similar to the action in walking, the water forcing open the after floats, a short pause then ensues, and the float falls by its own weight to its first position thereby preventing a retrograde motion of that float while the other is advancing in the same manner. To back water, the after floats are raised and the forward floats are lowered.

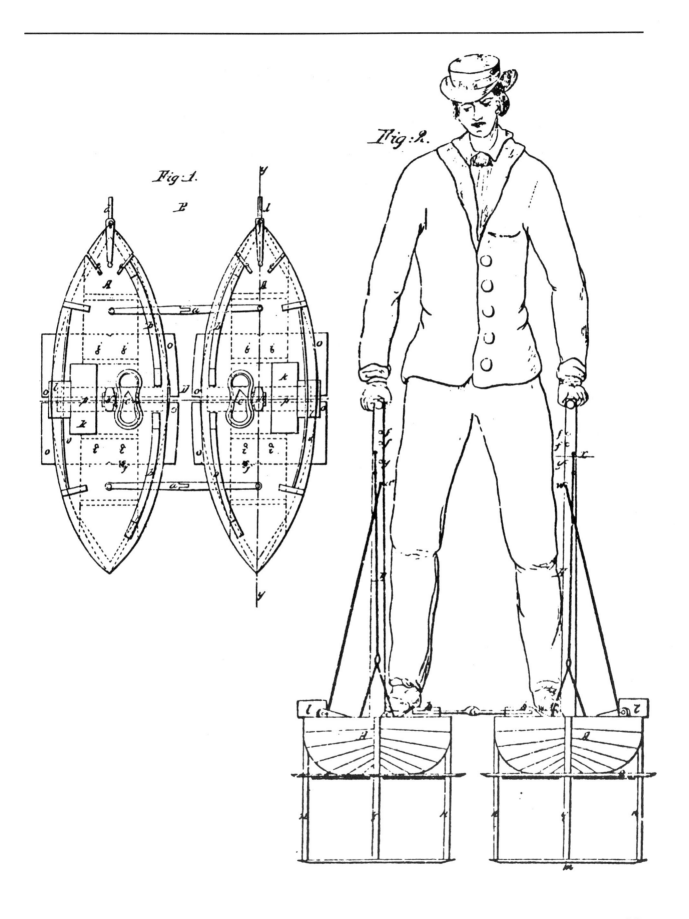

Submarine
Patent No. 581,213 (1897)
Simon Lake of Baltimore, Maryland

My invention [Fig. 1] relates to an improved submarine vessel, and has for its object, first, to provide novel means for sinking the vessel to the bottom of the water when it is at a state of rest or has no headway and for permitting the vessel to rise to the surface of the water; second, to provide means whereby the vessel is enabled to travel upon the bottom or bed of the water; third, to provide mechanism automatically controlled by the pressure of the water for submerging the vessel and maintaining it submerged at any desired or predetermined depth when underway; fourth, to provide means for automatically maintaining the vessel on a level keel irrespective of the disposal or shifting of the weights in the vessel; fifth, to provide novel means for affording ready ingress and egress from and to the vessel when submerged; and, lastly, to improve the construction generally and render more safe and certain the operation of submarine vessels. . . .

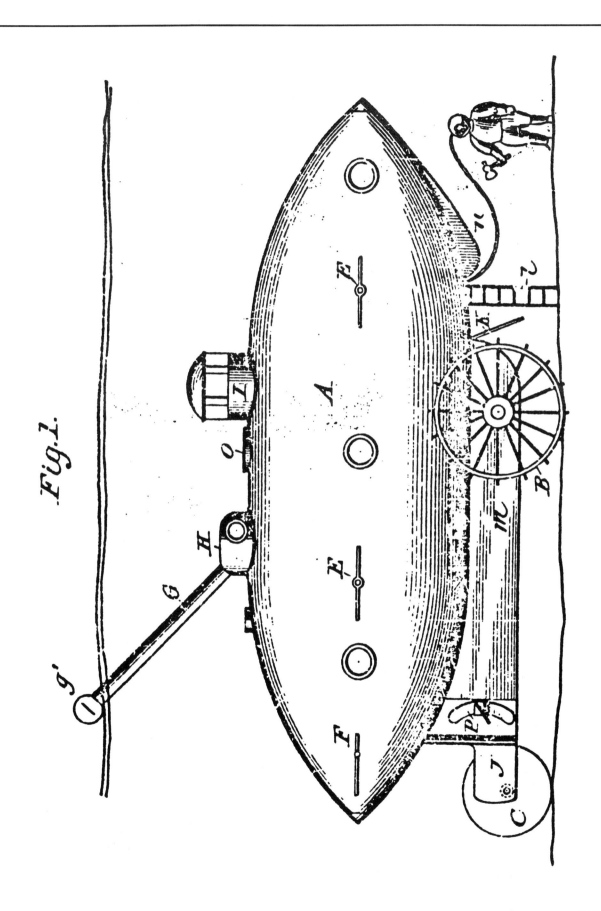

Balloon Birds
Patent No. 363,037 (1887)
Charles Richard Edouard Wolfe of Paris, France

By this present invention the mechanical motor and propelling and guiding arrangements are replaced by a living motor or motors taken from the flying classes of birds—such as, for example, one or more eagles, vultures, condors, &c. By means of suitable arrangements [see Figs. 1 and 4] all the qualities and powers given by nature to these most perfect kinds of birds may be completely utilized.

The corsets or harnesses p have forms and dimensions appropriate to the bodies of the birds chosen, such as eagles, vultures, condors, &c.

As the balloon floats in the air the man placed on the floor d can easily cause the cross $k\,k'$ to turn by means of the hand-wheel m, and, with the cross, the birds x, so as to utilize their flight in the direction of the axis of the balloon, or in any other direction he desires. On the other hand, by working the rollers r, r' he can direct the flight of the birds upward or downward. The result of these arrangements is that the flight of the harnessed birds must produce the motion and direction of the balloon desired by the conductor.

It may be observed that the birds have only to fly, the direction of their flight being changed by the conductor quite independently of their own will.

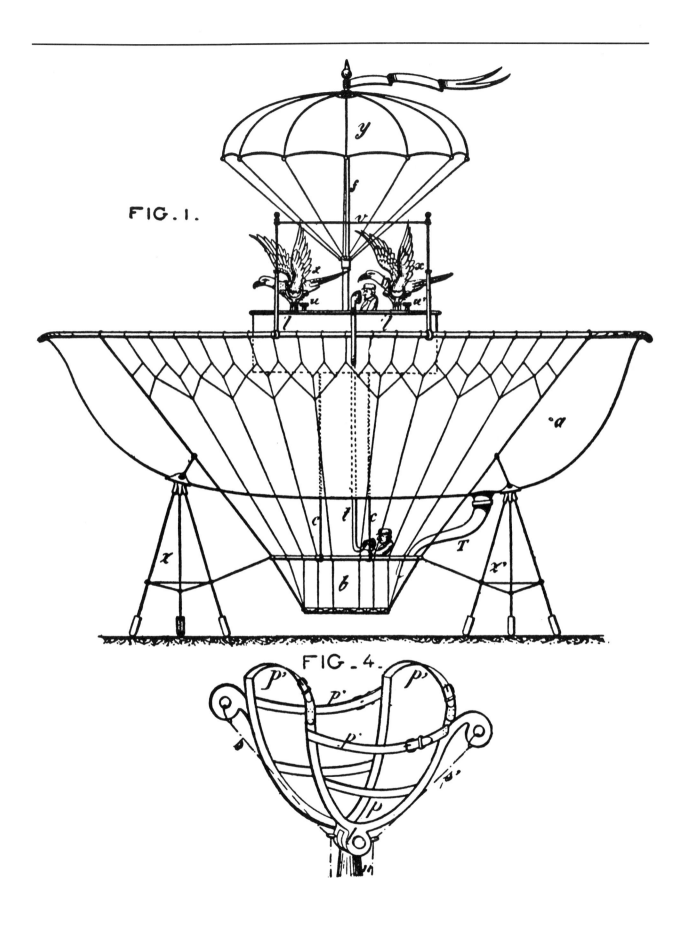

Dirigible
Patent No. 621,195 (1899)
Ferdinand Graf Zeppelin of Stuttgart, Germany

This invention relates to a navigable balloon which is characterized essentially in that it is provided with a number of motors arranged separately from each other. In this manner it is possible to give the balloon or buoyant part of the apparatus, which receives the gas and is preferably cylindrical with rounded ends, a smaller diameter in proportion to the driving power developed by the motors and to correspondingly reduce the air resistance. A navigable balloon or air craft of this kind can be combined with several other balloons or air crafts in such a manner that the foremost craft contains the driving-gear, while the others serve for the reception of the goods or load to be carried.

In the accompanying drawings, Figure 1 is a side elevation of my improved navigable balloon or air craft. Fig. 2 is a front elevation thereof....

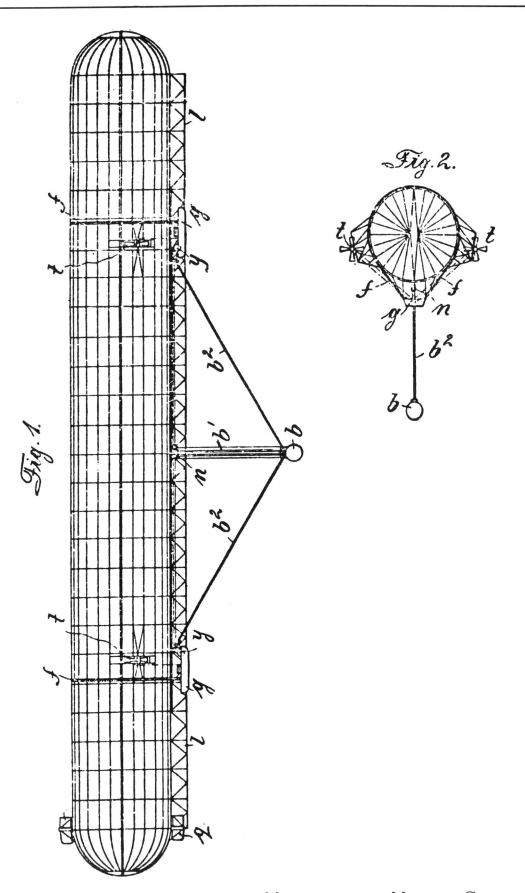

Airplane
Patent No. 821,393 (1906)
Orville and Wilbur Wright of Dayton, Ohio

Our invention relates to that class of flying-machines in which the weight is sustained by the reactions resulting when one or more aeroplanes are moved through the air edge-wise at a small angle of incidence, either by the application of mechanical power or by the utilization of the force of gravity.

The objects of our invention are to provide means for maintaining or restoring the equilibrium or lateral balance of the apparatus, to provide means for guiding the machine both vertically and horizontally, and to provide a structure combining lightness, strength, convenience of construction, and certain other advantages.

In flying-machines of the character to which this invention relates the apparatus is supported in the air by reason of the contact between the air and the under surface of one or more aeroplanes, the contact-surface being presented at a small angle of incidence to the air. The relative movements of the air and aeroplane may be derived from the motion of the air in the form of wind blowing in the direction opposite to that in which the apparatus is traveling or by a combined downward and forward movement of the machine, as in starting from an elevated position or by combination of these two things, and in either case the operation is that of a soaring-machine, while power applied to the machine to propel it positively forward will cause the air to support the machine in a similar manner. In either case owing to the varying conditions to be met there are numerous disturbing forces which tend to shift the machine from the position which it should occupy to obtain the desired results. It is the chief object of our invention to provide means for remedying this difficulty. [See Fig. 1.]

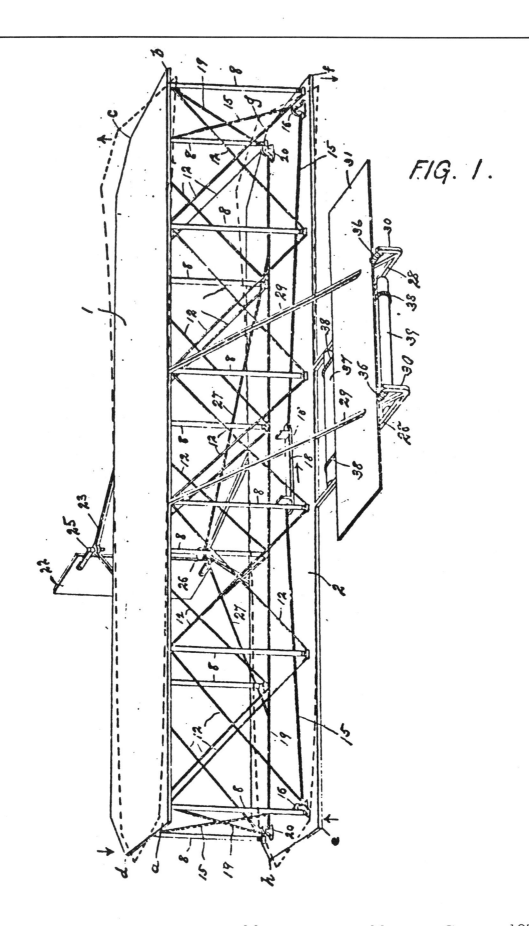

FIG. 1.

NINE
Eureka!

Some inventions have rhyme, some reason, and some simply defy description. To which Muse was Italian sculptor Benvenuto Cellini listening when he invented ball bearings in 1543? What kind of genius propelled Englishman John Walker to make the first friction matches in 1826? And whence the inspiration for sandpaper (Switzerland, fifteenth century), interchangeable parts (United States, 1802), or adhesive postage stamps (Great Britain, 1840)?

The impetus and ingenuity required for any number of other inventions cannot be easily conjured or categorized. Think about the fire

> He had been eight years upon a project for extracting sunbeams from cucumbers, which were to be put into phials hermetically sealed, and let out to warm the air in raw inclement summers.
> *Jonathan Swift,*
> Gulliver's Travels

extinguisher (1816), vulcanized rubber (1841), the gyroscope (1852), dynamite (1886), plastics (1868), aluminum making (1886), the slot machine (1889), the Geiger counter (1910), the supermarket (1930). At the time they were invented, they may well have been dismissed by many as the ideas of crackpots.

Then ask why some of the other inventions shown here never caught on.

No fair using hindsight.

Revolving Dinner Table
Patent No. 55,677 (1866)
W. L. Lance of Plymouth, Pennsylvania

The nature of my improvement [see Figs. 1 and 2] consists in providing a revolving or moving serving-table, *b,* driven by steam or other power, with either two or three rings or continuous tables, with shelves so loaded with viands, to move at the rate of fifteen or twenty feet per minute, or to pass before each guest at such speed as to exhibit before each guest the entire bill of fare once per minute.

All persons at this table are put upon an equality and free to act for themselves, and these shelves so arranged as not only to contain the full bill of fare, and that kept hot by lamp or otherwise, but also to contain all the necessary dishes, knives, forks, spoons, glasses, &c., and also so arranged as to carry the dishes that have been used off into the pantry *P,* behind the screen, where they are removed by the servant stationed at that point for that purpose, and where also are the persons stationed to supply and replenish the revolving or moving table *b,* with shelves *c d e f.* The carver and his assistants are also stationed behind the screen, which we here term "pantry," *P,* to supply continually the revolving or moving table *b* and shelves *c d e f.* The dishes, after the guest has finished with them, are put upon the lower shelf or table *b,* which is hid from view by means of a lid or curtain.

All the servants that we require in the use of this moving-table is one upon the outside and one upon the inside, except those required in the pantry to put away the last dishes of each guest and brush off the crumbs and adjust the chair. This would be the requirements of a table that would seat, say, one hundred and fifty persons.

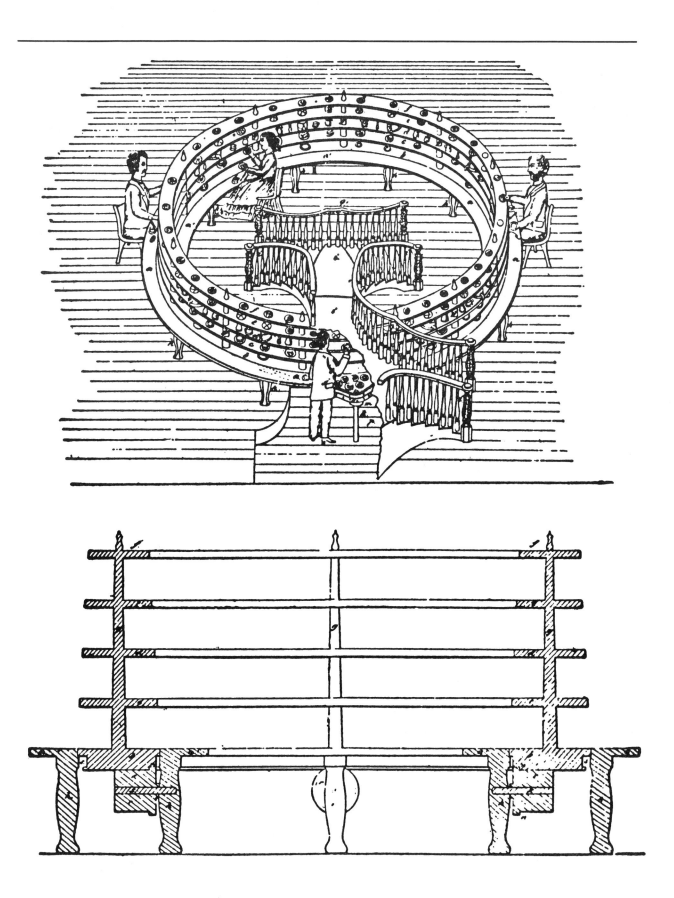

Mousetraps and Muffling Cups 201

Scrapbook
Patent No. 140,245 (1873)
Samuel Clemens (Mark Twain) of Hartford, Connecticut

Be it known that I, SAMUEL L. CLEMENS, of Hartford, in the county of Hartford and in the State of Connecticut, have invented certain new and useful Improvements in Scrap-Books.

[In Figs. 1 and 2] *A* and *B* represent two scrap-books of any desired dimensions, and made, as far as material, binding, &c., is concerned, in any of the known and usual ways. The leaves of which the book *A* is composed are entirely covered, on one or both sides, with mucilage or other suitable adhesive substance, while the leaves of which the book *B* is composed have the mucilage or adhesive substance applied only at intervals, as represented in Fig. 1.

In either case the scrap-book, so to say, self-pasting, as it is only necessary to moisten so much of the leaf as will contain the piece to be pasted in, and place such piece thereon, when it will stick to the leaf.

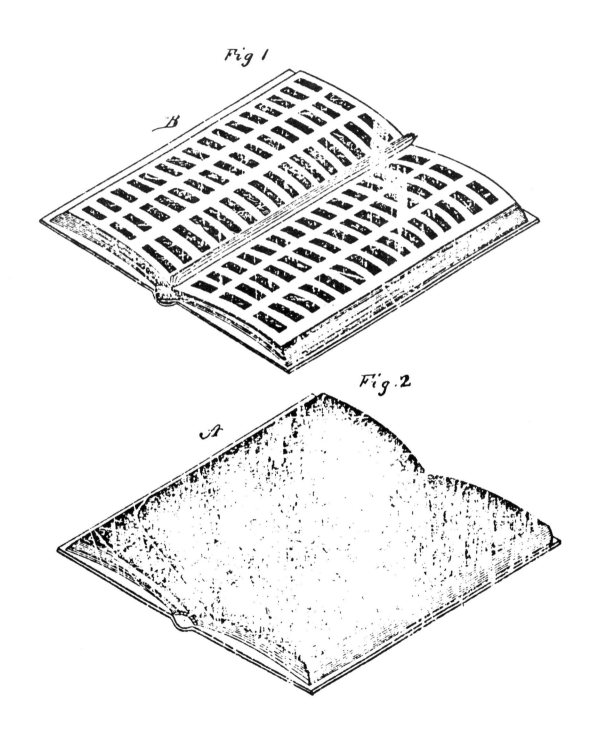

Short-Range Parachute
Patent No. 221,855 (1879)
Benjamin B. Oppenheimer of Trenton, Tennessee

This invention relates to an improved fire-escape or safety device, by which a person may safely jump out of the window of a burning building from any height, and land, without injury and without the least damage, on the ground.

Referring to [Fig. 1], A represents a head-piece, constructed in the nature of parachute, and made of soft or waxed cloth, awning-cloth, or other suitable fabric. The parachute is about four or five feet in diameter, stiffened by a suitable frame, and attached by a leather strap or other fastening, in reliable manner, to the head, neck, or arms.

In connection with the head-piece or parachute applied to the upper part of the body are used overshoes B, with elastic soles or pads C, of suitable thickness, that take up the sudden shock on arriving on the ground.

The parachute serves for the purpose of buoying the body in the air after the person has leaped from the window of the burning building, while the padded shoes secure the safe landing on the ground.

Escape Suspenders
Patent No. 323,416 (1885)
George C. Hale of Kansas City, Missouri

My invention [Figs. 1–3] relates to improvements in suspenders, having for its object to provide a suspender with a cord so secured thereto or formed therewith as to constitute a part of the same, and to be readily and easily detached therefrom, whereby, in the event of a person being confined to a burning building and having all of the usual means of escape cut off, the cords can be disengaged and lowered to the ground to receive a rope, and thus enable the person to effect his escape.

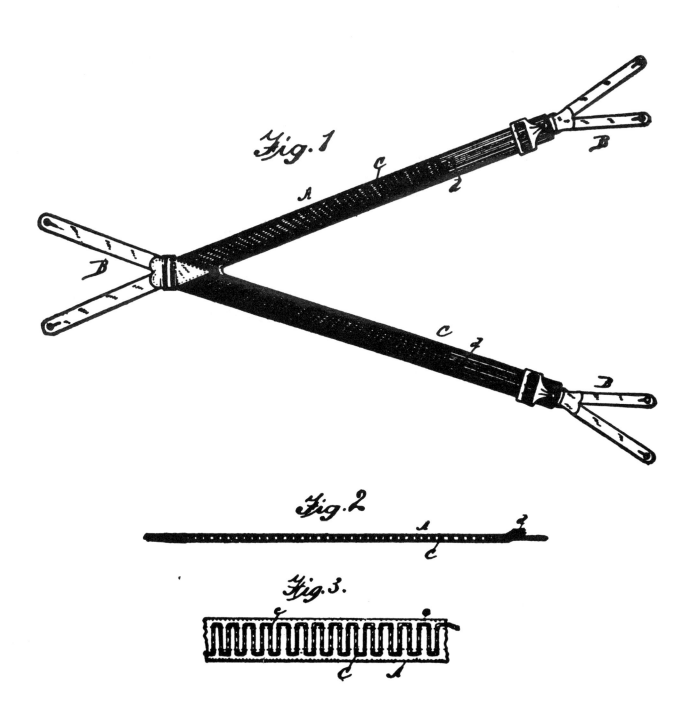

Chewing-Gum Locket
Patent No. 395,515 (1889)
Christopher Robertson of Somerville, Tennessee

The object of my invention [Figs. 1–4] is to provide a locket of novel form and construction for holding with safety, cleanliness, and convenience for use chewing-gum, confections, or medicines, and which may be carried in the pocket or otherwise attached to the person, as lockets are ordinarily worn.

As the lining B is made of a non-corrosive material, any of which may be employed without departing from my invention, the saliva of the mouth or other substance held within the locket will not act upon it chemically, and a case of any preferred material may thus be used. Chewing-gum may thus be carried conveniently upon the person, and is not left around carelessly to become dirty or to fall in the hands of persons to whom it does not belong, and be used by ulcerous or diseased mouths, by which infection would be communicated by subsequent use to the owner.

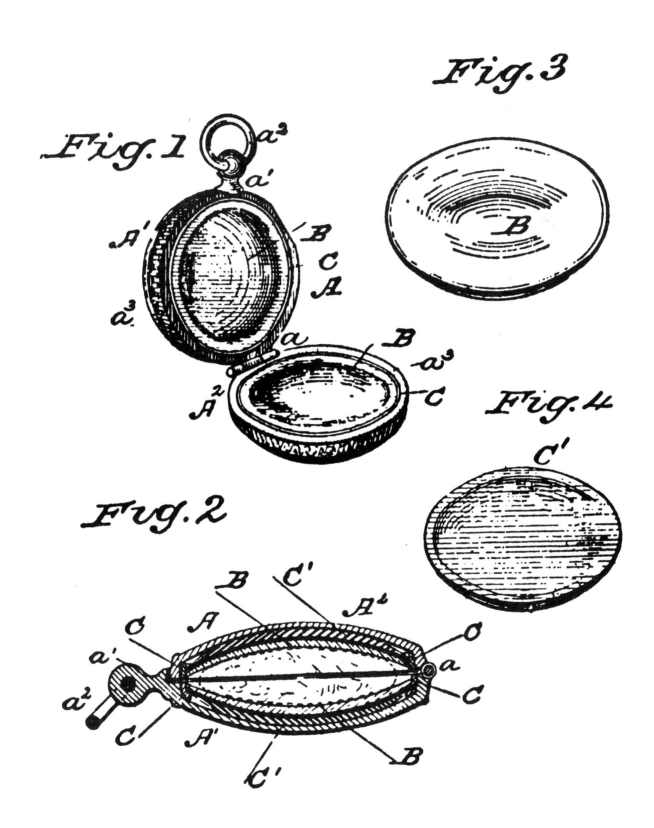

Body Preservation and Display
Patent No. 748,284 (1903)
Joseph Karwowski of Herkimer, New York

This invention has for its object the provision of a means whereby a corpse may be hermetically incased within a block of transparent glass, whereby being effectually excluded from the air the corpse will be maintained for an indefinite period in a perfect and life-like condition, so that it will be prevented from decay and will at all times present a life-like appearance [Figs. 1–3].

In carrying out my process I first surround the corpse I with a thick layer 2 of sodium silicate or water-glass. After the corpse has been thus inclosed within the layer of water-glass it is allowed to remain for a short time within a compartment or chamber having a dry heated temperature, which will serve to evaporate the water from this incasing layer, after which molten glass is applied to the desired thickness. This outer layer of glass may be molded into a rectangular form 3, as shown in Fig. 2 of the drawings, or, if preferred, cylindrical or other forms may be substituted for the rectangular block which I have illustrated. In Fig. 3 I have shown the head only of the corpse as incased within the transparent block of glass, it being at once evident that the head alone may be preserved in this manner, if preferred.

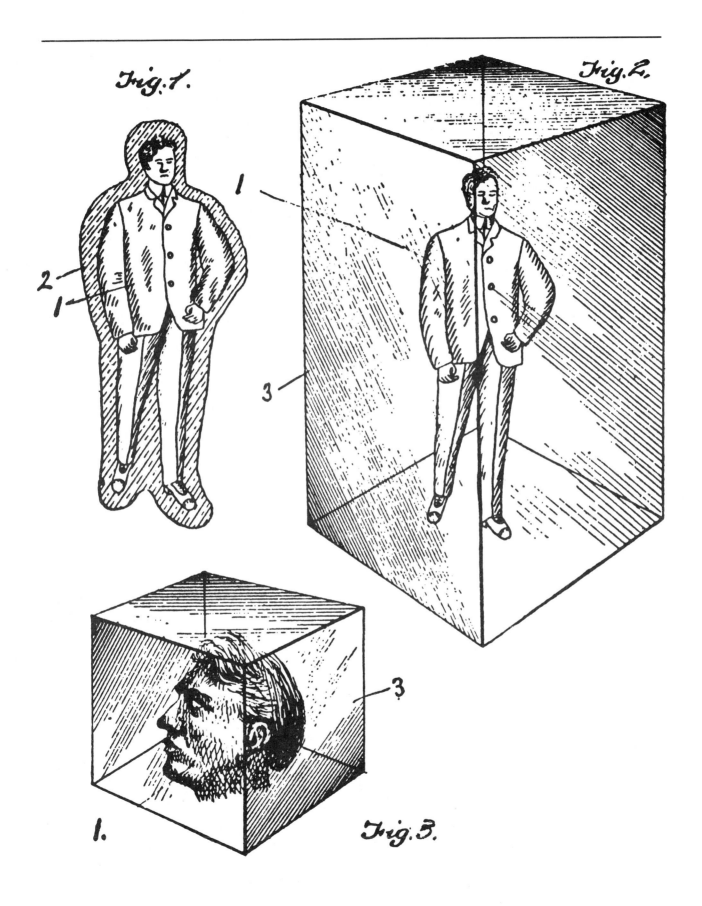

MOUSETRAPS AND MUFFLING CUPS

Improved Paper Clip
Patent No. 3,057,027 (1962)
Erling P. Bugge of Los Angeles, California

The present invention relates generally to fasteners and more particularly to an improved paper clip.

An object of my invention is to provide a paper clip having a greatly improved clamping action whereby to firmly and releasably secure papers together and also to more effectively resist lateral or longitudinal sliding movement of the clip on a clamped array of papers.

Another important object of the invention is to provide a paper clip of improved clamping efficiency but without any increase in the danger of tearing any of the clamped papers during removal of the clip, and as a matter of fact, greatly reducing the danger of such tearing.

It is also an object of the invention to provide a paper clip of the character described having portions adapted to positively inhibit the danger of any tearing of the clamped papers upon removal of the clip, such portions also being adapted to improve the clamping efficiency of the clip.

Yet another object of the invention is to provide a very attractive paper clip of this character having all its portions arranged in a substantially common plane and to achieve improved efficiency without any increase of dimension whereby the volume and weight of a package of my improved clips is no greater than in the case of conventional clips.

As presently constructed [Figs. 1–8], the preferred embodiment of my paper clip is made of a single length of resilient wire or like material. The clip comprises two clamping portions, the free ends of which can be separated to admit the papers to be clipped together between the clamping portions. These clamping arms or portions are of unequal length and the shorter of these is designated generally by the numeral 10 and the longer is designated generally by the numeral 11.

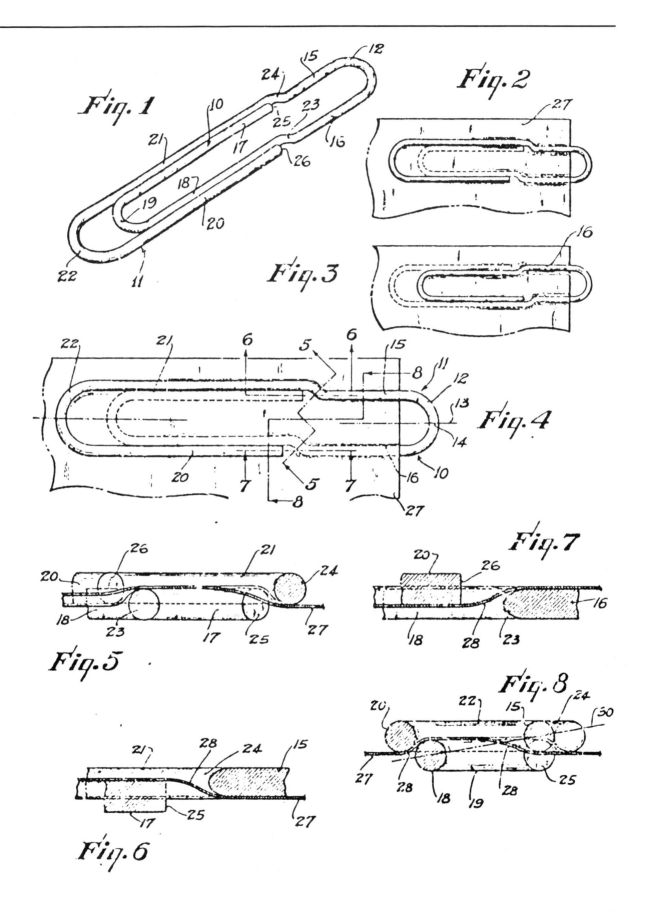

Improved Frisbee
Patent No. 3,359,678 (1967)
Edward E. Headrick of La Canada, California
(assignor to Wham-O Manufacturing Company)

This invention relates to aerodynamics toys to be thrown through the air and in particular to flying saucers for use in throwing games.

Over the past several years toys resembling saucers have become quite popular as throwing implements. In the usual embodiment the implement is made of a plastic material in a saucer shape with a rim located around the edge of the saucer, the rim having a somewhat greater thickness than the saucer portion of the implement. The rim curves downwardly from the saucer and has a configuration such that the implement when viewed in elevation approximates the shape of an airfoil.

The present invention provides an improved version of this well-known flying saucer. In this invention, means located on the convex side of the flying saucer [Figs. 1–3] are provided for interrupting the smooth flow of air over this surface. In aerodynamics this action is described as "spoiling" the air flow and the means by which this is accomplished are described as "spoilers." As applied to the present invention, this disruption of air flow is thought to create a turbulent unseparated boundary layer over the convex side of the saucer and to result in a reduction of drag especially in high-speed flight and an increase in stability while in flight. This means that a novice thrower can learn to throw the flying saucer more rapidly, that more expert throws will result with less experience, that better accuracy can be achieved and that a reduction in the skill required to use the saucer is made possible.

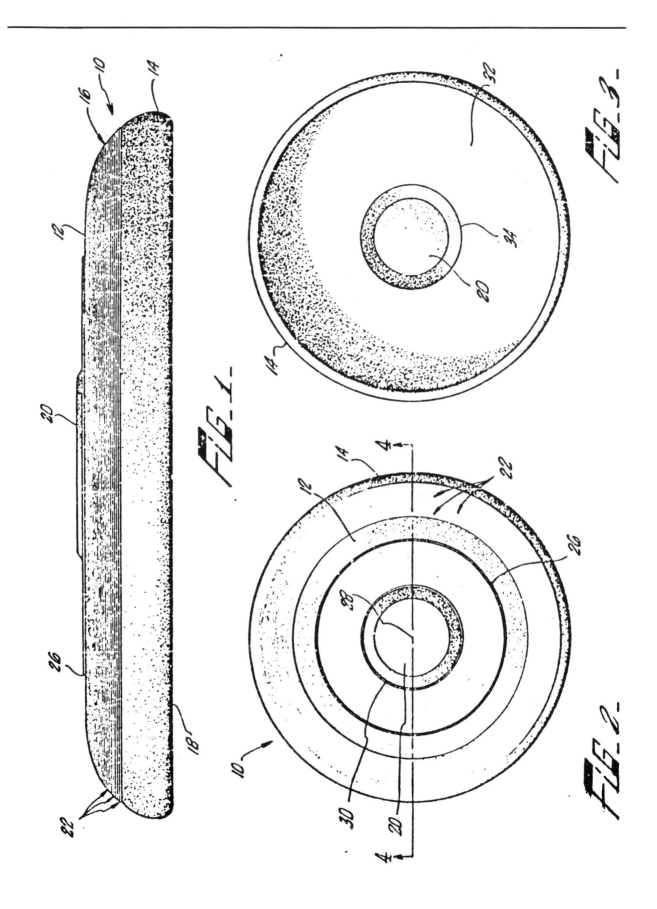

Lipstick Dispenser
Patent No. 3,934,598 (1976)
Thomas H. Hayes of Westport, Connecticut, and
Efrem Ostrowsky of Highland Park, Illinois

In the past a large number of cosmetic or lipstick holders and applicators have been proposed and produced. One class or prior device of this type employed a flexible push-pull strip or band which was fastened at one end to the cup and adjacent its other end to a finger piece which extended outwardly through a slot in the wall of the casing.

There were a number of disadvantages and drawbacks in these prior devices. In virtually all constructions heretofore proposed, the cases were either wholly, or at least in part, of square or rectangular cross section. This was due to the fact that the actuator strips employed therein were of appreciable width whereby their side edges extended along widely spaced, longitudinal portions of the casing. Such rectangular dispensers were difficult to manipulate and use, since they could not be easily twirled in the manner of a cylinder, in the hand of the user during application. It is well recognized that such rolling or twirling movement is desirable to facilitate the application of lipstick wax, for example.

The above drawbacks and disadvantages of prior cosmetic holders or lipstick devices are obviated by the present invention. A flexible push-pull activator strip located in the casing [see Fig. 5] has one end connected to the cup to activate the same, and has at its other end a concave-convex closure member. The cup, push-pull strip and closure member are guided and controlled by an internal curved guide member, all in such a manner that the casing can be completely round while at the same time the push-pull strip, when operated by a finger piece protruding from the casing, can advance or retract the cup and its product, and simultaneously automatically open or close the casing at its open end, depending on the location of the cup and its product.

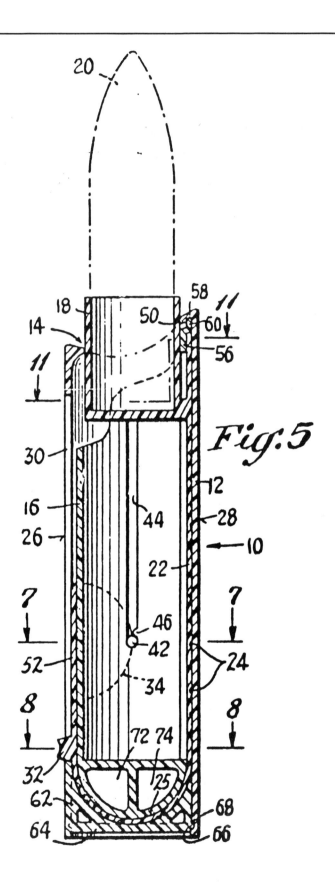

Muffling Cup
Patent No. 4,325,178 (1983)
Monya Scully of New York, New York

This invention is directed to providing a sound muffling cup into which an enraged person can shout to release tension while avoiding disturbing other persons. The cup [Fig. 1] comprises an elongated body portion having a substantially closed end wall at one end, and brim shaped to provide a mouthpiece at the other end. The body portion is of a size such that the fingers can be at least partially wrapped around the body portion for holding of the cup, and the mouthpiece is of a size and contour such that it can be placed over the mouth with substantially the entire brim contacting the skin along a generally elliptical line spaced from the lips and on the mouth side of the nose and chin.

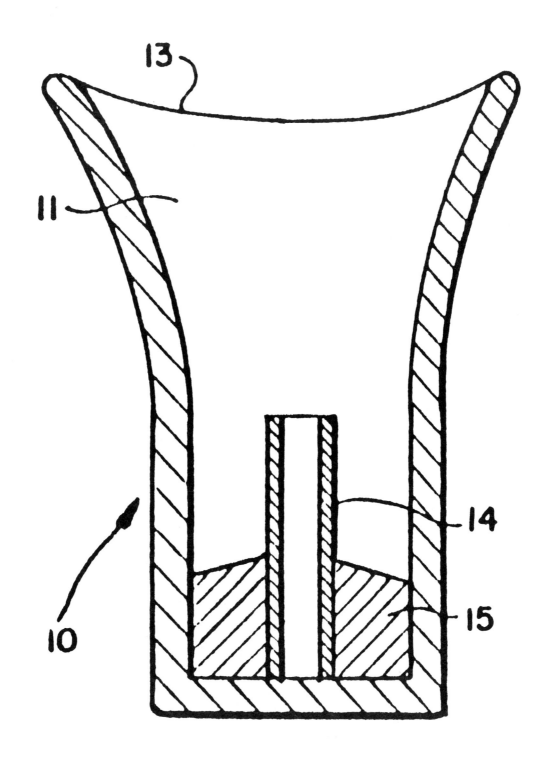